Oxford Resources for Cambridge

exam success

Cambridge IGCSE®

MATHEMATICS

Core & Extended

Revision Guide

Second Edition

Ian Bettison
Mathew Taylor

OXFORD
UNIVERSITY PRESS

Great Clarendon Street, Oxford, OX2 6DP, United Kingdom

Oxford University Press is a department of the University of Oxford.
It furthers the University's objective of excellence in research,
scholarship, and education by publishing worldwide. Oxford is a
registered trade mark of Oxford University Press in the UK and in
certain other countries.

© Ian Bettison and Mathew Taylor 2025

The moral rights of the authors have been asserted.

First published in 2025.

All rights reserved. No part of this publication may be reproduced, stored
in a retrieval system, transmitted, used for text and data mining, or used
for training artificial intelligence, in any form or by any means, without
the prior permission in writing of Oxford University Press, or as expressly
permitted by law, by licence or under terms agreed with the appropriate
reprographics rights organization. Enquiries concerning reproduction
outside the scope of the above should be sent to the Rights Department,
Oxford University Press, at the address above.

You must not circulate this work in any other form and you must impose
this same condition on any acquirer

British Library Cataloguing in Publication Data
Data available

978-1-382-05650-2

10 9 8 7 6 5 4 3 2 1

The manufacturing process conforms to the environmental regulations
of the country of origin.

Printed in Great Britain by Bell & Bain Ltd., Glasgow

The manufacturer's authorised representative in the EU for product
safety is Oxford University Press España S.A. of el Parque Empresarial
San Fernando de Henares, Avenida de Castilla, 2 - 28830 Madrid
(www.oup.es/en).

Acknowledgements

This Exam Success Revision Guide refers to the Cambridge IGCSE® Maths
syllabus published by Cambridge International Education.

This work has been developed independently from and is not endorsed
by or otherwise connected with Cambridge International Education.
IGCSE® is the registered trademark of Cambridge International Education.

The publisher would like to thank the following for permissions to use
copyright material:

Cover illustrations: James L. Amos / Corbis Documentary / Getty Images

Artwork by Straive, Aptara Inc., and Oxford University Press.

Links to third party websites are provided by Oxford in good faith and for
information only. Oxford disclaims any responsibility for the materials
contained in any third party website referenced in this work.

Contents

Introduction 4
   How to use this book 4
   How you will be assessed 8

1 Number 11
   Raise your grade 50

2 Algebra and graphs 51
   Raise your grade 95

3 Coordinate geometry 98
   Raise your grade 110

4 Geometry 111
   Raise your grade 139

5 Mensuration 140
   Raise your grade 153

6 Trigonometry 154
   Raise your grade 171

7 Transformations and vectors 174
   Raise your grade 186

8 Probability 188
   Raise your grade 201

9 Statistics 204
   Raise your grade 221

> **Answers:** find the numerical answers to all questions along with worked solutions and commentaries to all 'Raise your grade' exam-style questions in your digital book.

Introduction

How to use this book

Fully matched to the latest Cambridge assessment criteria, this in-depth Exam Success Revision Guide brings clarity and focus to exam preparation with detailed and practical guidance on raising attainment in IGCSE® Mathematics.

This Exam Success Revision Guide:

- is **fully matched** to the latest Cambridge IGCSE® syllabus
- includes a comprehensive list of **syllabus objectives** at the start of each chapter where you can build a record of your revision as well as **Key skills** features within the chapters to guide you through your revision and **Recap** features to review the key information
- provides exam-style questions at the end of each chapter to equip you to **Raise your grade**. These questions have fully worked solutions with commentaries as part of the online resources
- will guide you through answering exam questions with extensive use of **Worked examples** with **Exam tips**
- will help you to avoid common mistakes with **Watch out!** features.

This Exam Success Revision Guide has been designed to maximise exam potential. The key features which will help you include:

- **Your revision checklist** table with a list of the objectives covered in each section at the start of each unit to focus your learning, monitor progress and build a record of your revision. Objectives that are for the **Extended** syllabus only are shown in **bold** and with an Ⓔ icon.

Your revision checklist			
Core/Ⓔ **Extended** syllabus	☹	😐	🙂
1.1	○	○	○
1.2	○	○	○

- **Recap**: use these to review the key information.

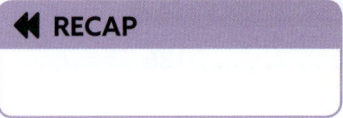

- **Apply**: short activities to help you to remember some of the key facts and ideas.

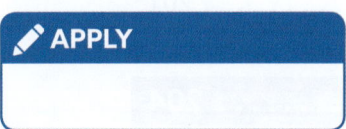

- **Key skills**: each section contains several key skills, to guide you through your revision, that you must master in order to succeed.

Introduction

- **Worked examples**: every section is rich in Worked examples which guide you step-by-step through answering exam-style questions.

  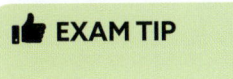

- **Exam tips**: alongside each Worked example are Exam tips which will help you to answer the question and set out your working correctly.

- **Watch out!**: these boxes point out key misconceptions and how to avoid them.

- **Questions**: at the end of each section there are questions to help you to practise the key skills.

  ? QUESTIONS

- **Raise your grade**: these sample exam-style questions appear at the end of each chapter for you to attempt.

  > **Answers**: find the numerical answers to all questions along with worked solutions and commentaries to all 'Raise your grade' exam-style questions in your digital book.

- **Calculator and non-calculator:** questions and worked examples marked with the icon allow the use of a calculator. You should not use a calculator unless you see this icon.

You could also create a revision planner like the one below to plan your revision timetable.

Monday	Tuesday	Wednesday	Thursday	Friday	Saturday	Sunday

Introduction

'Extended only' content

This book is intended for use by candidates studying both the Core and Extended syllabuses.

- **Your revision checklist** will indicate Extended objectives using an ⓔ icon.
- The syllabus objectives at the start of each chapter indicate **Extended only** content in **bold**.
- Within each chapter, content found in the Extended syllabus but not in the Core syllabus is introduced with a green 'Extended' heading and green shaded background.
- **Raise your grade** questions for the Extended syllabus are also clearly identified by the green 'Extended' heading and green shaded background.

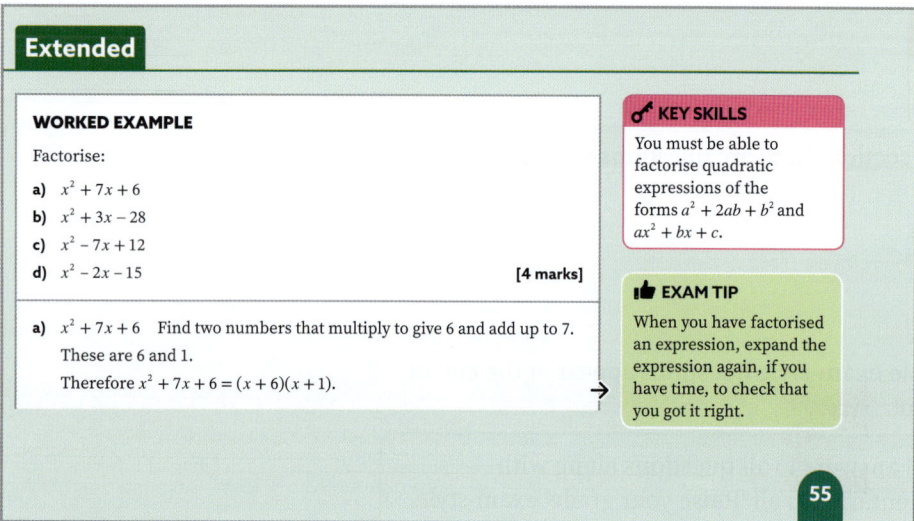

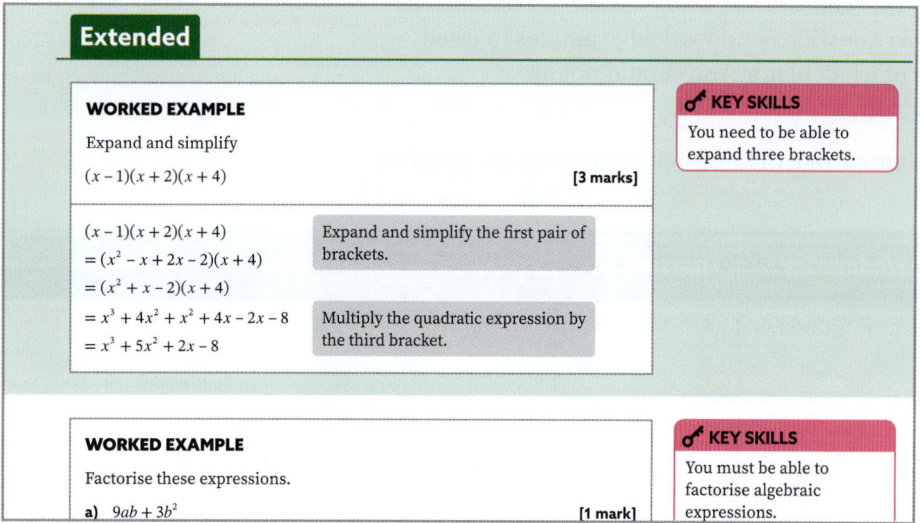

Introduction

Aiming for success

This Exam Success Revision Guide will help you to understand what is in the Cambridge IGCSE® Mathematics exam and to improve your key skills to achieve higher grades.

When you take your exam, you will sit two compulsory papers. Work through the chapters in order to help you to become more familiar with the demands of each section of the syllabus and to achieve greater success in the exam.

> **Answers**: use the full worked solutions and commentaries for each 'Raise your grade' question. These can be found in your digital book.

↑ Raise your grade

7. If a train travels for 50 km at a speed of 40 km h^{-1}, then travels a further 60 km at a speed of 30 km h^{-1}, find:
   a) the total time taken [3 marks]
   b) the average speed at which the train travelled, correct to 3 significant figures. [3 marks]

Oxford Resources for Cambridge — exam success — Cambridge IGCSE® MATHEMATICS — Revision Guide Core & Extended

a) Speed = distance ÷ time

   Time = distance ÷ speed

Write down the formula for speed and rearrange it.

   t_1 = 50 ÷ 40 = 1.25

   t_2 = 60 ÷ 30 = 2

Calculate the times for the two different parts of the journey.

   Total time taken = 1.25 + 2 = 3.25 hours (3 hours, 15 minutes)

Calculate the total time.

b) Average speed = total distance ÷ total time

Write down the formula for average speed.

   Total distance = 50 + 60 = 110 km

Calculate the total distance.

   Average speed = 110 ÷ 3.25 = 33.8 km h^{-1}

*Divide your total distance by your answer to part **a)** and round it to 3 significant figures.*

Each question in the 'Raise your grade' section is mapped to the relevant section in the chapter, so that you can go back and revise the key ideas if you get stuck.

How you will be assessed

The Cambridge IGCSE® Mathematics qualification (syllabus 0580) is examined at either Core tier (expected IGCSE® grades C to G) or at Extended tier (expected IGCSE® grades A* to C).

The assessment for each tier is the same. Candidates sit one short-answer paper and one paper with structured questions. The following tables give the details of each paper.

Core candidates take:

Paper 1 (Core)	1 hour 30 minutes
Non-calculator paper	
80 marks (50% of total marks awarded)	
Short-answer questions	
Questions will be based on the Core syllabus	
Externally assessed	

Extended candidates take:

Paper 2 (Extended)	2 hours
Non-calculator paper	
100 marks (50% of total marks awarded)	
Short-answer questions	
Questions will be based on the Extended syllabus	
Externally assessed	

and

Paper 3 (Core)	1 hour 30 minutes
Calculator paper	
80 marks (50% of total marks awarded)	
Short-answer questions	
Questions will be based on the Core syllabus	
Externally assessed	

and

Paper 4 (Extended)	2 hours
Calculator paper	
100 marks (50% of total marks awarded)	
Short-answer questions	
Questions will be based on the Extended syllabus	
Externally assessed	

- Paper 1 and Paper 2 do **not** allow the use of a calculator.
- Candidates should have a scientific calculator for Paper 3 and Paper 4.
- Three significant figures will be required in answers (or one decimal place for answers in degrees) except where otherwise stated.
- Candidates should use the value of π from their calculator or the value of 3.142.

Questions on the Core papers will be based only on the Core syllabus.

Questions on the Extended papers will be based on the Extended syllabus (which includes all of the Core syllabus content).

Advice and guidance

Before the exams:

- ✓ Check the dates of the exams
- ✓ Plan your revision carefully. Make a timetable that includes each section of the syllabus
- ✓ Focus on the areas that you have found difficult in the past
- ✓ Practise skills first, then applications to exam-style questions
- ✓ Create mind-maps, flash cards or revision posters if you think they will help
- ✓ Understand the command words in each question

Command word	What it means
Calculate	work out from given facts, figures or information, generally using a calculator
Construct	make an accurate drawing
Describe	state the points of a topic/give characteristics and main features
Determine	establish with certainty
Explain	set out purposes or reasons/make the relationships between things evident/provide why and/or how and support with relevant evidence
Give	produce an answer from a given source or recall/memory
Plot	mark point(s) on a graph
Show (that)	provide structured evidence that leads to a given result
Sketch	make a simple freehand drawing showing the key features
Work out	calculate from given facts, figures or information with or without the use of a calculator
Write	give an answer in a specific form
Write down	give an answer without significant working

- ✓ Use mark schemes and worked solutions to assess yourself
- ✓ Make sure that you have all of your equipment for the exam. You need a scientific calculator, pens, pencils, a ruler, a protractor, and a pair of compasses

During the exams:

✓ Plan your time. Allow one minute per mark so that you have some time at the end to check your answers

✓ If you can't do a question, move on quickly and come back to it

✓ Read the questions carefully and identify the command word(s)

✓ Set out your working clearly so it can be followed by the examiner

✓ Make sure that your final answer is clearly indicated

✓ Check that you have included the units, if appropriate

✓ Check that you have rounded your answer to a suitable or specified degree of accuracy

✓ Check that your answer is sensible

✓ Do not round intermediate values in working—this may lead to rounding errors in your final answer

✓ Make sure that you use correct mathematical terminology when giving reasons

✓ Go back and attempt questions you missed out at the end

✓ Leave time at the end for checking through all of your answers again

1 Number

Your revision checklist

Tick these circles to build a record of your revision.

Core/ **E** **Extended** syllabus

		☹ 😐 🙂
1.1	Identify and use: natural numbers, integers (positive, zero and negative), prime numbers, square numbers, cube numbers, common factors, common multiples, rational and irrational numbers, reciprocals.	○ ○ ○
1.2	Understand and use set language, notation and Venn diagrams to describe sets.	○ ○ ○
	E Represent relationships between sets.	○ ○ ○
1.3	Calculate with the following: squares, square roots, cubes, cube roots, other powers and roots of numbers.	○ ○ ○
1.4	Use the language and notation of the following in appropriate contexts: proper fractions, improper fractions, mixed numbers, decimals, percentages.	○ ○ ○
	Recognise equivalence and convert between these forms.	○ ○ ○
	E Write fractions in their simplest form.	○ ○ ○
1.5	Order quantities by magnitude and demonstrate familiarity with the symbols $=, \neq, >, <, \geq$ and $\leq$.	○ ○ ○
1.6	Use the four operations for calculations with integers, fractions and decimals, including correct ordering of operations and use of brackets.	○ ○ ○
1.7	Understand and use indices (positive, zero and negative).	○ ○ ○
	E Understand and use fractional indices.	○ ○ ○
	Understand and use the rules of indices.	○ ○ ○
1.8	Use the standard form $A \times 10^n$ where n is a positive or negative integer and $1 \leq A < 10$.	○ ○ ○
	Convert numbers into and out of standard form.	○ ○ ○
	Calculate with values in standard form.	○ ○ ○
1.9	Round values to a specified degree of accuracy.	○ ○ ○
	Make estimates for calculations involving numbers, quantities and measurements.	○ ○ ○
	Round answers to a reasonable degree of accuracy in the context of a given problem.	○ ○ ○
1.10	Give upper and lower bounds for data rounded to a specified accuracy.	○ ○ ○
	E Find upper and lower bounds of the results of calculations which have used data rounded to a specified accuracy.	○ ○ ○
1.11	Understand and use ratio and proportion to: give ratios in their simplest form, divide a quantity in a given ratio, use proportional reasoning and ratios in context.	○ ○ ○
1.12	Use common measures of rate.	○ ○ ○
	Apply other measures of rate.	○ ○ ○
	Solve problems involving average speed.	○ ○ ○
1.13	Calculate a given percentage of a quantity.	○ ○ ○
	Express one quantity as a percentage of another.	○ ○ ○
	Calculate percentage increase or decrease.	○ ○ ○
	Calculate with simple and compound interest.	○ ○ ○
	E Calculate using reverse percentages.	○ ○ ○
1.14	Use a calculator efficiently.	○ ○ ○
	Enter values appropriately on a calculator.	○ ○ ○
	Interpret the calculator display appropriately.	○ ○ ○
1.15	Calculate with time: seconds (s), minutes (min), hours (h), days, weeks, months, years, including the relationship between units.	○ ○ ○
	Calculate times in terms of the 24-hour and 12-hour clock.	○ ○ ○
	Read clocks and timetables.	○ ○ ○
1.16	Calculate with money.	○ ○ ○
	Convert from one currency to another.	○ ○ ○
1.17	**E** Use exponential growth and decay.	○ ○ ○
1.18	**E** Understand and use surds, including simplifying expressions.	○ ○ ○
	E Rationalise the denominator.	○ ○ ○

1 Number

1.1 Number properties

> **YOU NEED TO:**
> - Identify and use: natural numbers, integers (positive, zero and negative), prime numbers, square numbers, cube numbers, common factors, common multiples, rational and irrational numbers, reciprocals.

Natural numbers are the numbers you use to count. So the natural numbers are 1, 2, 3, 4, …

Integers are 'whole numbers'. They can be positive or negative (with zero in between). So the integers are the numbers … −3, −2, −1, 0, 1, 2, 3, …

Positive integers are the numbers 1, 2, 3, 4, …

Negative integers are the numbers −1, −2, −3, −4, …

Prime numbers have only two (different) factors (i.e. 1 and the number itself). So 1 is not a prime number. The prime numbers are 2, 3, 5, 7, 11, 13, 17, 19, … (The number of primes is infinite.)

Square numbers
$1^2 = 1$, $2^2 = 4$, $3^2 = 9$, … So the numbers 1, 4, 9, … are square numbers.

$1^2 = 1 \times 1 = 1$ $2^2 = 2 \times 2 = 4$ $3^2 = 3 \times 3 = 9$

Cube numbers
$1^3 = 1$, $2^3 = 8$, $3^3 = 27$, … So the numbers 1, 8, 27, … are cube numbers.

    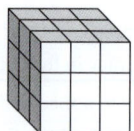

$1^3 = 1 \times 1 \times 1 = 1$ $2^3 = 2 \times 2 \times 2 = 8$ $3^3 = 3 \times 3 \times 3 = 27$

> ✏️ **APPLY**
>
> Make a list of:
> - the first 15 prime numbers
> - the first 15 square numbers
> - the first 8 cube numbers.

WORKED EXAMPLE

Express 504 as the product of prime factors. **[2 marks]**

Divide 504 by the smallest possible prime number, in this case 2.

Continue until you have only prime numbers in your tree.

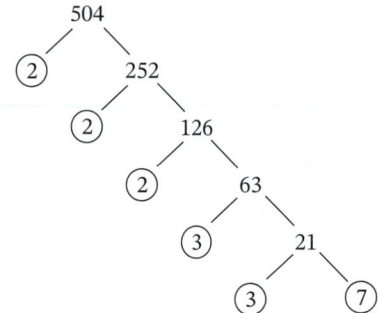

Hence $504 = 2^3 \times 3^2 \times 7$

> 🗝️ **KEY SKILLS**
>
> You need to be able to write any number as the **product of prime factors**.

> 👍 **EXAM TIP**
>
> Use a factor tree to help you. Give your answer using index notation.

12

3 is a common factor of 9 and 12 since 3 is a factor of both 9 and 12.

30 is a common multiple of 6 and 15 since 30 is a multiple of both 6 and 15.

> **WORKED EXAMPLE**
>
> Find the highest common factor and lowest common multiple of 60 and 504. **[4 marks]**
>
> ---
>
> $60 = 2^2 \times 3 \times 5$
>
> $504 = 2^3 \times 3^2 \times 7$
>
> The highest common factor is the product of the prime factors *common* to both numbers, in this case, 2^2 and 3.
>
> HCF $= 2^2 \times 3 = 12$
>
> The lowest common multiple is the product of the largest power of each prime number that appears in either number, in this case, 2^3, 3^2, 5 and 7.
>
> LCM $= 2^3 \times 3^2 \times 5 \times 7 = 2520$

🔑 KEY SKILLS

You need to be able to work out the **Highest Common Factor (HCF)** and the **Lowest Common Multiple (LCM)** of two numbers.

👍 EXAM TIP

First, write both numbers as the product of prime factors. 2^2 is common to both numbers, but 2^3 is not. Similarly, 3 is common to both numbers, but 3^2 is not.

A **rational number** is a number that can be expressed in the form $\frac{p}{q}$ where p and q are whole numbers.

- All decimals that recur are rational. For example, $0.\dot{3} = 0.333333...$ is rational since $0.\dot{3} = \frac{1}{3}$
- All decimals that terminate (i.e. which end) are rational. For example, 0.625 is a rational number since 0.625 can be written as $\frac{625}{1000} = \frac{5}{8}$

An **irrational number** is a number that cannot be expressed in the form $\frac{p}{q}$, where p and q are integers. For example, $\pi, \sqrt{2}$ and $\sqrt[3]{7}$ are all irrational numbers.

A **real number** is any rational or irrational number that can be represented on a number line.

👁 WATCH OUT!

$\sqrt{2\frac{1}{4}}$ does not look rational but it is, since

$\sqrt{2\frac{1}{4}} = \sqrt{\frac{9}{4}} = \frac{3}{2}$

1 Number

WORKED EXAMPLE

$\sqrt{4}$ $\sqrt{15}$ $\sqrt{25}$ $\sqrt{36}$ $\sqrt{144}$

From the list above, write down:

a) an odd prime number
b) two factors of 42
c) an irrational number
d) a multiple of 4. [4 marks]

a) $\sqrt{25} = 5$, which is an odd prime number
b) $\sqrt{4} = 2$ and $\sqrt{36} = 6$, and both 2 and 6 are factors of 42
c) $\sqrt{15}$ is an irrational number
d) $\sqrt{144} = 12$, which is a multiple of 4

EXAM TIP
Look for square numbers in the list and work out the square roots of these first.

The **reciprocal** of a number is 'one over' the number. For example, the reciprocal of 4 is $\frac{1}{4}$ and the reciprocal of n is $\frac{1}{n}$. If you are asked to find the reciprocal of a fraction, you just 'flip it over'. For example, the reciprocal of $\frac{2}{3}$ is $\frac{3}{2}$.

? QUESTIONS

1. Which of these numbers is **not** a rational number?

    $\frac{3}{5}$ $\sqrt{7}$ 0.6 $-1\frac{2}{3}$ $\sqrt{25}$

2. An integer n is such that $80 \leq n < 90$.
    Write down a value of n that is:
    a) a multiple of both 3 and 4
    b) a prime number
    c) a factor of 1700.

3. Find:
    a) a prime number which is a factor of 49
    b) an even prime number.

4. a) Write down all the factors of 21.
    b) Write down all the factors of 28.
    c) Hence find the HCF of 21 and 28.

5. $\sqrt{9}$ $\sqrt{15}$ $\sqrt{49}$ $\sqrt{16}$ $\sqrt{64}$ $\sqrt{121}$
    From the list above, write down:
    a) a prime number less than 10
    b) a factor of 22
    c) a power of 2
    d) an irrational number.

6. Write each number as the product of prime factors (e.g. $72 = 2^3 \times 3^2$).
    a) 30
    b) 24
    c) 18
    d) 28
    e) 105
    f) 64

7. Use question **6** to find the highest common factor of each pair of numbers.
    a) 30 and 24
    b) 28 and 64
    c) 18 and 105

8. Use question **6** to find the lowest common multiple of each pair of numbers.
    a) 30 and 64
    b) 24 and 105
    c) 28 and 18

9. Write down the reciprocal of each of these numbers.
    a) 5
    b) $\frac{1}{3}$
    c) $\frac{4}{5}$
    d) $\frac{17}{13}$

To **Raise your grade** now try question 3 on page 50

1.2 Sets

> **YOU NEED TO:**
> - Understand and use set language, notation and Venn diagrams to describe sets.
> - 🇪 Represent relationships between sets.

A **set** is a collection of items. These may be numbers, people, letters, etc.

You use 'curly' brackets to represent sets.

For example, the set P, of prime numbers less than or equal to 10, can be represented as $P = \{2, 3, 5, 7\}$.

The number of **elements** in set P is denoted by n(P), so, in this case, n(P) = 4.

A **Venn diagram** can be used to represent sets.

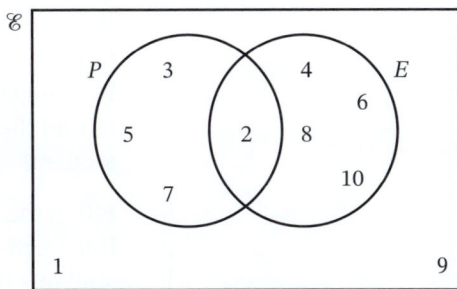

In the Venn diagram above, P is the set of prime numbers less than or equal to 10 and E is the set of even numbers less than or equal to 10.

The symbol $\mathscr{E}$ is the symbol for the **universal set**. This is the set of all things that are being considered at the time, in this case, the set of integers from 1 to 10.

The element '2' lies inside both circles. This region of the Venn diagram is known as the **intersection** and contains all of the elements that are in **both** sets.

> ⏪ **RECAP**
>
> The symbol for intersection is ∩. In the Venn diagram above:
>
> $P \cap E = \{2\}$
> n($P \cap E$) = 1

Elements that lie in either P or E or both lie in the **union** of P and E.

> ⏪ **RECAP**
>
> The symbol for union is ∪. In the Venn diagram above:
>
> $P \cup E = \{2, 3, 4, 5, 6, 7, 8, 10\}$
> n($P \cup E$) = 8

Elements that do not lie in P are called the **complement** of P.

> ⏪ **RECAP**
>
> The symbol for the complement of set P is P'.
>
> In the Venn diagram above:
>
> $P' = \{1, 4, 6, 8, 9, 10\}$
> n(P') = 6

1 Number

WORKED EXAMPLE

In a class of 33 students, 20 like chess, 12 like draughts and 5 like neither.

a) Represent this information in a Venn diagram. **[3 marks]**

b) How many students like only one of chess or draughts? **[1 mark]**

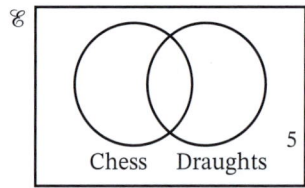

a) There are 33 − 5 = 28 students who like either chess or draughts or both.

There are 20 who like chess and 12 who like draughts.

20 + 12 − 28 = 4

So 4 students like both chess and draughts.

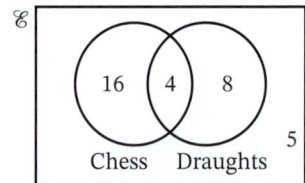

b) The number of students who like only one of chess or draughts is 16 + 8 = 24

> **EXAM TIP**
>
> Draw a Venn diagram with two circles overlapping and add in the 5 students who like neither game outside both circles.
>
> Add up the number of students who like chess and who like draughts and then subtract the number of students who like at least one of the games to find the number of students who like both.
>
> Fill in the 4 in the intersection. 20 − 4 = 16 students like chess only. 12 − 4 = 8 students like draughts only.

WORKED EXAMPLE

Draw Venn diagrams and shade the region(s) that represent:

a) $A' \cup B$

b) $A' \cap B$ **[2 marks]**

a) $A' \cup B$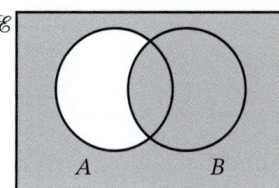

b) $A' \cap B$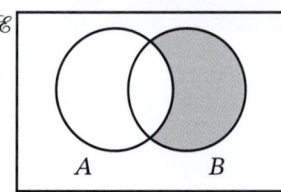

> **KEY SKILLS**
>
> You need to be able to shade Venn diagrams to represent different situations.

> **EXAM TIP**
>
> Imagine shading the region A' and the region B separately.
>
> A'  B
>
> The union is any region that is shaded in *either* diagram.
>
> The intersection is any region that is shaded in *both* of the separate diagrams.

Extended

◀◀ RECAP

You need to know several other symbols and notation for sets.

∈ means 'is an element of'

∉ means 'is not an element of'

∅ means the empty set

$A \subseteq B$ means 'A is a subset of B' (and A can equal B)

$A \not\subseteq B$ means 'A is not a subset of B'

WORKED EXAMPLE

Draw Venn diagrams and shade the region(s) that represent:

a) $(B \cup C) \cap A$

b) $(A \cap C) \cup B'$ [2 marks]

a) $(B \cup C) \cap A$

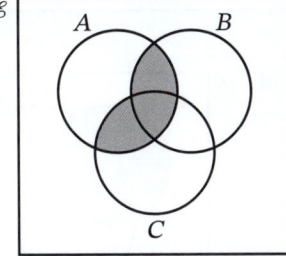

b) $(A \cap C) \cup B'$

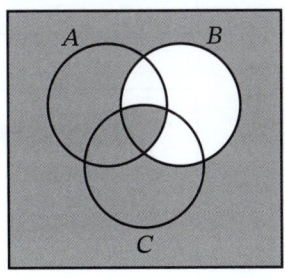

✏️ APPLY

$A = \{$Integers from 1 to 10$\}$

$B = \{$Odd numbers between 0 and 20$\}$

$C = \{2, 4, 6, 8, 10\}$

Which of the following statements are true and which are false?

- $2 \in C$
- $4 \in B$
- $8 \notin A$
- $14 \notin B$
- $A \subseteq B$
- $B \cap C = \emptyset$

🔑 KEY SKILLS

You need to be able to work with Venn diagrams containing three circles.

👍 EXAM TIP

Look for the regions that lie in B or C and **also** in A.

👍 EXAM TIP

The only region in B that is shaded is the part that belongs to all three sets.

1 Number

? QUESTIONS

1. In a group of 100 students, 70 enjoy Maths, 50 enjoy French and 20 enjoy neither.

   a) Draw a Venn diagram showing this information.

   b) Use your diagram to find the number of students who enjoy both subjects.

2. On an athletics day, 150 athletes take part. 60 are in the 100 metres race, 50 are in the 200 metres race and 80 are in neither.

   a) Draw a Venn diagram showing this information.

   b) Use the diagram to find the number of athletes who ran in only one race.

3. In a shop there were 120 customers on a certain day. Of these, 60 paid using notes, 30 paid using coins and 50 paid using cards. There were no customers who paid using both cards and cash.

   a) Draw a Venn diagram showing this information.

   b) Use your diagram to find the number of customers who used both notes and coins.

Extended

4. Use set notation to define the shaded regions:

   a)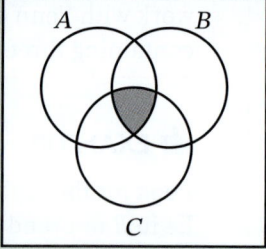

   b)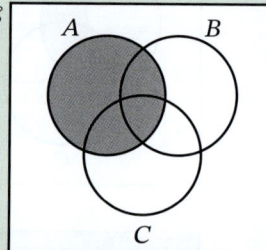

   c)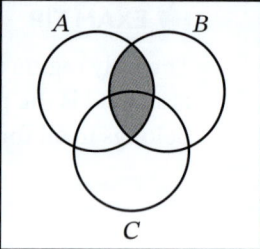

   d)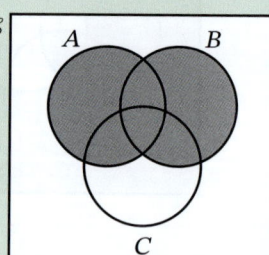

1.3 Powers and roots

> **YOU NEED TO:**
> - Calculate with the following:
>   squares, square roots, cubes, cube roots, other powers and roots of numbers.

The following examples show keys that feature on most calculators.

The **square** of a number n is $n \times n = n^2$, so the square of 5 is $5^2 = 25$

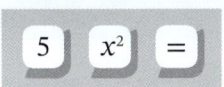 on some calculators

The **cube** of a number n is $n \times n \times n = n^3$, so the cube of 2 is $2^3 = 8$

 on some calculators

$25 = 5^2$
5 is the **square root** of 25.
The square root of n is represented by $\sqrt{n}$. So $\sqrt{25} = 5$

 on some calculators

$8 = 2^3$
2 is the **cube root** of 8.
The cube root of n is represented by $\sqrt[3]{n}$. So $\sqrt[3]{8} = 2$

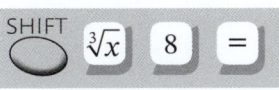 on some calculators

> **APPLY**
> Investigate how to enter powers and roots on your scientific calculator.

> **EXAM TIP**
> You will be expected to know the squares of numbers from 1 to 15 and their corresponding square roots, and the cubes of 1, 2, 3, 4, 5, and 10 and their corresponding cube roots.

> ◀◀ **RECAP**
> $2^4 = 16$, so $\sqrt[4]{16} = 2$
> $3^5 = 243$, so $\sqrt[5]{243} = 3$

WORKED EXAMPLE

A cube of side l metres has a volume of 30 cubic metres.

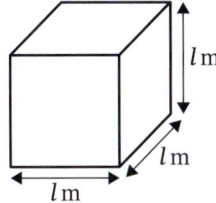

a) Calculate the value of l.
b) Calculate the area of one face of the cube. **[3 marks]**

---

a) $l = \sqrt[3]{30} = 3.107$
$= 3.11$ metres (to 3 s.f.)

b) $(3.107...)^2 = 9.654$
$= 9.65$ square metres (to 3 s.f.)

> 🗝 **KEY SKILLS**
> You need to be able to work with other powers and roots, for example, the power 4 and fourth roots.

> **EXAM TIP**
> Use your calculator to find the cube root of 30 since $l^3 = 30$.
>
> In part **b)**, use the *exact* answer from part **a)** rather than the rounded answer. This ensures no rounding errors are carried forwards.

19

1 Number

WORKED EXAMPLE

Use your calculator to work out $6^4 \div \sqrt[6]{1000}$. Round your answer to two decimal places. **[2 marks]**

$6^4 \div \sqrt[6]{1000} = 409.831$
$= 409.83$ (to 2 d.p.)

> **EXAM TIP**
>
> In a question like this, you will normally get one mark for evidence that you have entered the correct calculation and one mark for giving your answer to the correct level of accuracy (as specified in the question).

QUESTIONS

1. Find l (to 3 s.f.) given that the volume of the cube is $50\,\text{cm}^3$.

   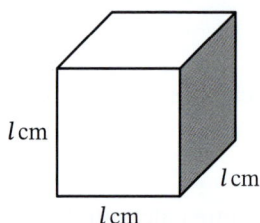

2. Calculate the value of each expression (to 3 s.f. where necessary).
   a) 8^2
   b) 11^3
   c) 7.1^2
   d) $\sqrt{17}$
   e) $\sqrt[3]{-100}$
   f) $\sqrt[3]{6859}$

3. Find the side length of a square whose area is $49\,\text{cm}^2$.

4. Find the side length of a cube whose volume is $216\,\text{cm}^3$.

5. Write the following in ascending order (smallest first).
   a) $0.6^3 \quad \sqrt[3]{0.6} \quad \sqrt{0.6} \quad 0.6 \quad 0.6^2$
   b) $6^3 \quad \sqrt[3]{6} \quad \sqrt{6} \quad 6 \quad 6^2$

6. Calculate the value of each expression, rounding to 3 s.f. where appropriate.
   a) $3^2 \times \sqrt[4]{16}$
   b) $2^5 \div \sqrt[3]{8}$
   c) $5^4 \times \sqrt{7}$
   d) $4^6 \div \sqrt[5]{20}$

1.4 Fractions, decimals and percentages

> **YOU NEED TO:**
> - Use the language and notation of the following in appropriate contexts: proper fractions, improper fractions, mixed numbers, decimals, percentages.
> - Recognise equivalence and convert between these forms.
> - **E** Write fractions in their simplest form.

If a cake is split into five equal pieces and Joe eats one of the five pieces, then the amount he eats can be expressed:

- as a **fraction**, that is, $\frac{1}{5}$ of the whole cake
- as a **decimal**, that is, 0.2 of the whole cake
- as a **percentage**, that is, 20% of the whole cake.

These three expressions are equivalent.

WORKED EXAMPLE

Write $\frac{24}{30}$ in its simplest form. [2 marks]

$$\frac{24}{30} = \frac{24 \div 6}{30 \div 6} = \frac{4}{5}$$

KEY SKILLS

You need to be able to simplify fractions.

EXAM TIP

Divide both the numerator and denominator by 6, which is the highest common factor of 24 and 30.

WORKED EXAMPLE

a) Express 0.15 as:
   i) a fraction
   ii) a percentage. [3 marks]
b) Express $\frac{2}{5}$ as a decimal. [1 mark]

a) 0.15 is expressed as:
   i) $0.15 = \frac{15}{100} = \frac{3}{20}$
   ii) $0.15 \times 100\% = 15\%$.
b) $\frac{2}{5} = \frac{4}{10} = 0.4$

KEY SKILLS

You need to be able to convert between fractions, decimals and percentages.

EXAM TIP

Multiplying by 100% does not change the value of the number, since 100% = 1.

1 Number

Extended

⏮ RECAP

To convert a recurring decimal to a fraction, multiply the recurring decimal by the power of 10 equal to the number of recurring digits. For example, if there are 2 recurring digits, multiply by 100.

🗝 KEY SKILLS

You need to be able to convert from a recurring decimal to a fraction in its simplest form.

WORKED EXAMPLE

Convert $0.\dot{1}\dot{2}$ into a fraction in its simplest form. **[3 marks]**

$$x = 0.1212...$$
$$100x = 12.1212...$$
Let $100x - x = 12.1212... - 0.1212...$
$$99x = 12$$
$$x = \frac{12}{99} = \frac{4}{33}$$

👍 EXAM TIP

Multiply x by 100 since there are 2 recurring digits.

Subtract x from $100x$ and then simplify the resulting fraction, if possible.

❓ QUESTIONS

1. Simplify the following fractions.
   a) $\frac{10}{15}$ b) $\frac{18}{20}$
   c) $\frac{35}{40}$ d) $\frac{120}{150}$

2. Express 0.375 as:
   a) a fraction
   b) a percentage.

3. Express $\frac{3}{25}$ as:
   a) a decimal
   b) a percentage.

4. Express 55% as:
   a) a decimal
   b) a fraction.

5. 0.037   37%   0.307 $\frac{3}{10}$ $\frac{3}{100}$ 3.7%

   From the numbers listed above, write down:
   a) the smallest number
   b) the largest number
   c) the two numbers that are equal.

Extended

6. Convert the following recurring decimals to fractions.
   a) $0.\dot{2}$ b) $0.\dot{6}\dot{1}$ c) $0.1\dot{3}\dot{5}$ d) $0.\dot{6}1\dot{6}$

22

1.5 Ordering quantities

YOU NEED TO:

- Order quantities by magnitude and demonstrate familiarity with the symbols $=, \neq, >, <, \geq$ and $\leq$.

WORKED EXAMPLE

Place these numbers in ascending order.

$\frac{34}{50}$ $\frac{2}{3}$ 0.66   67% $\frac{13}{20}$ [2 marks]

In ascending order they are:

$\frac{13}{20}$ 0.66 $\frac{2}{3}$ 67% $\frac{34}{50}$

KEY SKILLS

You need to be able to compare the sizes of numbers given as fractions, decimals and percentages.

EXAM TIP

Write all of the numbers as decimals to compare them.

Remember to list them in order in their original forms.

WORKED EXAMPLE

$a = 5$

$b = -4$

Choose one of the symbols $=, <$ or $>$ to complete each of these statements.

a) $a + b \ldots b + a$

b) $a^2 \ldots b^2$

c) $a - b \ldots b - a$ [3 marks]

a) $5 + (-4) = 1$ and $(-4) + 5 = 1$

so $a + b = b + a$

b) $5^2 = 25$ and $(-4)^2 = 16$

so $a^2 > b^2$

c) $5 - (-4) = 9$ and $(-4) - 5 = -9$

so $a - b > b - a$

EXAM TIP

Work out the value of each expression first.

You can use the same symbol more than once.

QUESTIONS

1. Write these numbers in ascending order.

   $\frac{41}{50}$ 0.8 $\frac{21}{25}$ 81%   0.85

2. If $x = 9$ and $y = 11$, write the correct sign, $=, <$ or $>$, in these expressions.

   a) $x \ldots y$

   b) $y - x \ldots x - y$

   c) $12x \ldots y^2$

1 Number

1.6 The four operations

YOU NEED TO:

- Use the four operations for calculations with integers, fractions and decimals, including correct ordering of operations and use of brackets.

◀◀ RECAP

When adding or subtracting whole numbers or decimals, make sure you line up the digits according to their **place value**.

$$\begin{array}{r} \overset{1}{3647} \\ + 451 \\ \hline 4098 \end{array} \qquad \begin{array}{r} \overset{6}{6}\overset{12}{7}.\overset{1}{3}0 \\ - 15.61 \\ \hline 51.69 \end{array}$$

Check that you carry or borrow correctly.

◀◀ RECAP

When multiplying whole numbers, set out your working carefully as a column multiplication.

$$\begin{array}{r} 241 \\ \times\ 72 \\ \hline 482 \\ 16870 \\ \hline 17352 \end{array}$$

Make sure to add a '0' when you multiply by the 10s digit.

◀◀ RECAP

When multiplying decimals, ignore the decimal points in your long multiplication and then count the number of digits after the decimal points to give the correct final answer.

$$\begin{array}{r} 132 \\ \times\ 14 \\ \hline 528 \\ 1320 \\ \hline 1848 \end{array}$$

Hence $13.2 \times 1.4 = 18.48$ (two digits after the decimal point).

◀◀ RECAP

Division of whole numbers and decimals can be done using the 'bus stop' method.

$1446 \div 6$:
$$6 \overline{\smash{)}1^1 4\,^2 46} \rightarrow 241$$

$16.5 \div 11$:
$$11 \overline{\smash{)}16\,^5.5} \rightarrow 1.5$$

24

1.6 The four operations

WORKED EXAMPLE

Calculate:

a) $\dfrac{3}{4} + \dfrac{1}{5}$

b) $2\dfrac{1}{2} - 1\dfrac{2}{7}$

c) $2\dfrac{2}{3} \times 4\dfrac{1}{2}$

d) $\dfrac{6}{7} \div \dfrac{2}{5}$ [4 marks]

a) $\dfrac{3}{4} + \dfrac{1}{5} = \dfrac{15}{20} + \dfrac{4}{20} = \dfrac{19}{20}$

b) $2\dfrac{1}{2} - 1\dfrac{2}{7} = \dfrac{5}{2} - \dfrac{9}{7} = \dfrac{35}{14} - \dfrac{18}{14} = \dfrac{17}{14} = 1\dfrac{3}{14}$

c) $2\dfrac{2}{3} \times 4\dfrac{1}{2} = \dfrac{8}{3} \times \dfrac{9}{2} = \dfrac{72}{6} = 12$

d) $\dfrac{6}{7} \div \dfrac{2}{5} = \dfrac{6}{7} \times \dfrac{5}{2} = \dfrac{30}{14} = \dfrac{15}{7} = 2\dfrac{1}{7}$

> **◀◀ RECAP**
>
> Arithmetic with fractions:
> - When adding and subtracting fractions, use a common denominator.
> - When multiplying fractions, multiply the numerators and multiply the denominators.
> - When dividing fractions, flip the second one over and multiply.
> - If you have a mixed number, write the fraction as an improper fraction first.

BIDMAS stands for

Brackets – for example, (3 + 4)

Indices – for example, 2^3 or $\sqrt{3}$

Division – for example, 8 ÷ 2

Multiplication – for example, 3 × 4

Addition – for example, 5 + 2

Subtraction – for example, 7 − 2

> **🔑 KEY SKILLS**
>
> You need to be able to apply the correct order of operations when carrying out multi-step calculations.

WORKED EXAMPLE

Calculate $(3 + 4)^2 \times 3 + (5 + 7) \div 6 - 1$ [3 marks]

Using the B of BIDMAS gives $(3 + 4)^2 \times 3 + (5 + 7) \div 6 - 1$
$= 7^2 \times 3 + 12 \div 6 - 1$

Using the I of BIDMAS gives $7^2 \times 3 + 12 \div 6 - 1$
$= 49 \times 3 + 12 \div 6 - 1$

Using the D of BIDMAS gives $49 \times 3 + 12 \div 6 - 1$
$= 49 \times 3 + 2 - 1$

Using the M of BIDMAS gives $49 \times 3 + 2 - 1$
$= 147 + 2 - 1$

Using the A of BIDMAS gives $147 + 2 - 1 = 149 - 1$

Using the S of BIDMAS gives $149 - 1 = 148$

> **👍 EXAM TIP**
>
> Show each step of your working since there will be method marks available for certain key steps.

1 Number

You need to know how to use directed numbers in practical situations.

To add a positive number, or subtract a negative number, move to the right on a number line.

$(-4) + 6 = 2$

To subtract a positive number, or add a negative number, move to the left on a number line.

$5 + (-2) = 5 - 2 = 3$

Multiplying or dividing two negative numbers or two positive numbers gives a positive number.

Multiplying or dividing a negative number and a positive number gives a negative number.

$(-4) \times (-2) = 8 \qquad 3 \times (-5) = -15$

$(-10) \div (-2) = 5 \qquad (-6) \div 3 = -2$

> **WORKED EXAMPLE**
>
> In June, the average temperature in Moscow was 14 °C.
>
> In January, the average temperature was 25 °C lower than in June.
>
> What was the average temperature in Moscow in January? **[2 marks]**
>
> $14 - 25 = -11$ °C

🔑 KEY SKILLS

You need to be able to use **directed numbers** in practical situations.

👁 WATCH OUT!

You can use the rule 'two negatives make a positive' when multiplying or dividing.

When adding or subtracting, the same rule does not work so use the number line method instead.

👍 EXAM TIP

Sketch a number line if necessary.

❓ QUESTIONS

1. Calculate:
   a) $3752 + 681$
   b) $1563 - 789$
   c) 461×63
   d) $686 \div 7$

2. Calculate:
   a) $14.78 + 11.41$
   b) $19.31 - 7.9$
   c) 1.23×7.81
   d) $97.2 \div 12$

3. A wooden rod has length 150 cm. It is cut into pieces of length $1\frac{1}{4}$ cm.
   How many such pieces can be cut from the original rod?

4. A bottle of orange juice holds $2\frac{1}{2}$ litres of water. A glass holds $\frac{1}{8}$ litre.
   How many glasses can be filled from one bottle of orange juice?

5. a) Express 0.375 as a fraction.
   b) Express $\frac{3}{25}$ as a decimal.

6. Calculate:

   a) $3\frac{1}{2} \times 5\frac{1}{3}$

   b) $7\frac{3}{4} + 6\frac{2}{5}$

   c) $8\frac{5}{6} - 3\frac{3}{8}$

   d) $4\frac{1}{5} \div 2\frac{3}{7}$

7. Evaluate the following expressions.

   a) $(6-2) \times 7^2$

   b) $8 \times 2 - (4-1)^2$

   c) $(2^2 + 1) \div 5^2$

   d) $(4+7) \times (6^2 - 7^2)^2$

8. Add one set of brackets to each calculation to make it correct.

   a) $4 + 15 \div 5 - 2 \times 5 = 29$

   b) $3 + 2^2 \times 3 = 15$

   c) $5 + 3 \times 2 + 7 = 32$

9. The average temperature each month in Montreal is shown in the table.

Month	Jan	Feb	Mar	Apr	May	Jun	Jul	Aug	Sep	Oct	Nov	Dec
Temperature (°C)	−11	−9	−3	5	13	18	21	19	15	8	1	−7

   a) Find the difference between the highest and lowest average monthly temperatures.

   b) The average minimum temperature for December is usually 5 °C lower than the average December temperature. Find the average minimum temperature for December.

10. Calculate:

    a) $17 + (-3)$

    b) $(-4) \times (-2)$

    c) $(-7) - (-1)$

    d) $15 - (-4)$

    e) $(-18) \div (-3)$

    f) $(-40) \div 10$

11. A man's bank balance is −£345.20. He then withdraws £50. What is his new balance?

12. The water level on a gauge is at −20 cm, that is, 20 cm below the flood level. If the water rises by 25 cm what level does the gauge now show?

To **Raise your grade** now try question 1 on page 50

1 Number

1.7 Indices

YOU NEED TO:
- Understand and use indices (positive, zero and negative).
- **(E) Understand and use fractional indices.**
- Understand and use the rules of indices.

When you write $3 \times 3 \times 3 \times 3$ as 3^4 you are using index notation.

The number '3' is referred to as the **base** and the little number '4' is the **index**.

To multiply powers of the same base, add the indices.

$9^3 \times 9^4 = 9^7$

To divide powers of the same base, subtract the indices.

$2^9 \div 2^4 = 2^5$

To find a power of a power, multiply the indices.

$(5^3)^4 = 5^{12}$

◀◀ RECAP

The three laws of indices can be written in a more general form:

- $a^m \times a^n = a^{m+n}$
- $a^m \div a^n = a^{m-n}$
- $(a^m)^n = a^{mn}$

✏ APPLY

Match the expressions that have the same value. Do not use a calculator.

$5^6 \quad 5^2 \quad 5^4 \times 5^2 \quad 5^8 \quad 5 \times 5^2 \quad \dfrac{5^5}{5^3} \quad \dfrac{5 \times 5^8}{5^3} \quad 5^5 \times 5^3 \quad (5^4)^2 \quad 25^3 \quad 5^{11} \div 5^8$

WORKED EXAMPLE

Express these in the form a^n, where a is an integer.

a) $(11^3)^2 \times (11^2)^5$

b) $\dfrac{(7^3)^4 \times 7^8}{(7^2)^5}$

[4 marks]

a) $(11^3)^2 \times (11^2)^5 = 11^6 \times 11^{10} = 11^{16}$

b) $\dfrac{(7^3)^4 \times 7^8}{(7^2)^5} = \dfrac{7^{12} \times 7^8}{7^{10}} = \dfrac{7^{20}}{7^{10}} = 7^{10}$

🔑 KEY SKILLS

You need to be able to use the rules of indices to simplify expressions.

👁 WATCH OUT!

Remember to **subtract** the indices when dividing powers of a number; a common mistake is to divide them.

👍 EXAM TIP

Make sure you set out your working clearly and apply each rule in turn.

1.7 Indices

- $a^0 = 1$
- $a^{-1} = \dfrac{1}{a}$
- In general, $a^{-n} = \dfrac{1}{a^n}$
- $a^{\frac{1}{2}} = \sqrt{a}$
- In general, $a^{\frac{1}{n}} = \sqrt[n]{a}$
- In general, $a^{\frac{m}{n}} = \sqrt[n]{a^m} = \left(\sqrt[n]{a}\right)^m$

KEY SKILLS
You need to understand and be able to work with the zero index and negative indices.

WORKED EXAMPLE
Calculate:

a) 2^5 [1 mark]
b) 3^{-2} [1 mark]
c) 5^0 [1 mark]

a) $2^5 = 2 \times 2 \times 2 \times 2 \times 2 = 32$
b) $3^{-2} = \dfrac{1}{3^2} = \dfrac{1}{9}$
c) $5^0 = 1$

EXAM TIP
Remember that $a^{-b} = \dfrac{1}{a^b}$ and that $a^0 = 1$, as long as $a \neq 0$.

Extended

WORKED EXAMPLE
Evaluate these numbers.

a) $8^{\frac{4}{3}}$
b) $25^{-\frac{1}{2}}$
c) $\left(\sqrt{\dfrac{3}{2}}\right)^4$ [6 marks]

a) $8^{\frac{4}{3}} = \left(\sqrt[3]{8}\right)^4 = 2^4 = 16$
b) $25^{-\frac{1}{2}} = \dfrac{1}{\sqrt{25}} = \dfrac{1}{5}$
c) $\left(\sqrt{\dfrac{3}{2}}\right)^4 = \left(\left(\dfrac{3}{2}\right)^{\frac{1}{2}}\right)^4 = \left(\dfrac{3}{2}\right)^2 = \dfrac{9}{4}$

KEY SKILLS
You need to understand and work with fractional indices.

EXAM TIP
In part **a)**, apply the cube root first.

In part **b)**, remember that the negative in the power is 'one over'.

In part **c)**, convert the square root to a power of one-half and use the power of a power rule.

1 Number

WORKED EXAMPLE

If $\sqrt{128} = 2^k$, find the value of k. [2 marks]

$128 = 2^7$

$\sqrt{2^7} = (2^7)^{\frac{1}{2}} = 2^{\frac{7}{2}}$

Hence $k = \dfrac{7}{2}$

APPLY

You should learn the first eight powers of 2; try listing them now.

EXAM TIP

Although in theory it does not matter in which order you apply the different parts of an index (power, root or reciprocal), you may find that in practice it is easier to perform the root first to avoid having to find roots of very big numbers.

EXAM TIP

Use the rules of indices and make sure you actually answer the question by stating the value of k.

QUESTIONS

1. Calculate the values of these expressions.
   a) 2^6
   b) 3^4
   c) 5^3
   d) 11^2
   e) 2^{-3}
   f) 10^{-2}
   g) 19^0
   h) 13^2
   i) 4^{-3}
   j) $\left(\dfrac{1}{2}\right)^3$
   k) $\left(\dfrac{2}{3}\right)^2$
   l) $\left(\dfrac{5}{3}\right)^4$
   m) $\left(\dfrac{1}{2}\right)^{-2}$
   n) $\left(\dfrac{2}{5}\right)^{-3}$
   o) $\left(\dfrac{2}{5}\right)^{-3}$

2. Find x when:
   a) $32^x = 2$
   b) $81^x = 3$
   c) $125^x = 5$
   d) $49^x = 7$
   e) $121^x = 11$
   f) $27^x = 3$
   g) $243^x = 3$
   h) $256^x = 16$
   i) $3^x = \dfrac{1}{3}$
   j) $81^x = \dfrac{1}{3}$
   k) $125^x = \dfrac{1}{5}$
   l) $512^x = \dfrac{1}{2}$

3. Express these numbers in decimal form (to 3 s.f.).
   a) 0.2^6
   b) 0.3^{15}
   c) 0.15^7
   d) 0.22^8

4. Evaluate these expressions without using decimals.
   a) 3^{-4}
   b) 5^{-3}
   c) 7^{-2}
   d) $8^{\frac{1}{3}}$
   e) $16^{\frac{3}{4}}$
   f) $25^{\frac{3}{2}}$
   g) $\left(\dfrac{3}{4}\right)^2$
   h) $\left(\dfrac{27}{64}\right)^{\frac{1}{3}}$
   i) $\left(\dfrac{16}{625}\right)^{\frac{1}{4}}$

1.8 Standard form

YOU NEED TO:

- Use the standard form $A \times 10^n$ where n is a positive or negative integer and $1 \leq A < 10$.
- Convert numbers into and out of standard form.
- Calculate with values in standard form.

The mass of the Earth is about 5 974 200 000 000 000 000 000 000 kg, a very large number. The time taken for light to travel 1 km is about 0.000 003 335 56 seconds, a very small number. Standard form is useful when writing very large and very small numbers.

To write a number in standard form, express it as a number between 1 and 10 multiplied by the appropriate power of 10.

KEY SKILLS

You need to be able to convert numbers into standard form.

A is a number between 1 and 10 ($1 \leq A < 10$)

$A \times 10^n$

n is a whole number, positive for large numbers, negative for small numbers.

So the mass of the Earth = 5.9742×10^{24} kg

The time taken for light to travel 1 km = 3.33556×10^{-6} seconds

⏪ RECAP

To convert large numbers into standard form, move the digits to the right through the decimal point until only one non-zero digit remains. Count the number of places the digits have moved; this is n.

To convert small numbers into standard form, move the digits to the left through the decimal point until only one non-zero digit remains. Count the number of places the digits have moved; this is n, but remember that n is *negative*.

WORKED EXAMPLE

Write these numbers in standard form.

a) 3 723 000
b) 0.000 000 312 **[2 marks]**

a) 3.723×10^6
b) 3.12×10^{-7}

👍 EXAM TIP

For part **b)**, move the digits seven places to the left so that the decimal point comes after the first 3.

WORKED EXAMPLE

The speed of light is 3×10^8 m s^{-1}.

The distance from Earth to the Moon is 3.844×10^5 km.

Calculate, to three significant figures, the time it takes light reflected by the Moon to reach the Earth. **[3 marks]**

KEY SKILLS

You need to be able to calculate with numbers in standard form.

1 Number

speed = distance ÷ time, so time = distance ÷ speed

$$\frac{3.844 \times 10^8}{3 \times 10^8} = 1.28 \text{ seconds (to 3 s.f.)}$$

> **EXAM TIP**
>
> Convert 3.844×10^5 km into metres first by multiplying it by 1000 or 10^3.
>
> It is usually fine to give your answer as a normal number or in standard form, but read the question carefully to check.

✎ APPLY

Investigate how you can enter numbers in standard form into your scientific calculator.

❓ QUESTIONS

1. The distance from London to Beirut is approximately 3460 km. Express this number in standard form.

2. The distance from Auckland to Rio de Janeiro is approximately 12 260 km. Express this number in standard form.

3. The radius of the Earth is 6 378 100 m. Express this number in standard form.

4. The speed of light is given as $2.998 \times 10^8 \text{ m s}^{-1}$. Express this as an ordinary number.

5. Give the value of each of these expressions in standard form (to 3 s.f.).

   a) 2^{30} b) 3^{20} c) $\left(\frac{1}{2}\right)^6$

   d) $\left(\frac{1}{3}\right)^3$ e) $\sqrt{0.005}$ f) $\left(\frac{1}{11}\right)^2$

   g) 5^{13} h) $\sqrt{0.007}$ i) $\left(\frac{3}{4}\right)^{10}$

6. Give your answers to these calculations in standard form (to 3 s.f.).

   a) $(3.4 \times 10^7) \times (4.2 \times 10^5)$
   b) $(2.9 \times 10^{15}) \times (2.1 \times 10^7)$
   c) $(7.2 \times 10^4) \times (1.3 \times 10^{-1})$
   d) $(3.91 \times 10^{-5}) \div (2.35 \times 10^{-7})$
   e) $(9.21 \times 10^7) \div (2.31 \times 10^{-5})$
   f) $(1.21 \times 10^{-5}) \div (1.24 \times 10^9)$

7. The population of the world at the end of 1995 was 5.2×10^9 people.

   a) The population was projected to grow by 4% in 1996. Calculate the projected population at the end of 1996, giving your answer in standard form (to 2 s.f.).

   b) In fact, the population at the end of 1996 was 5.5×10^9. What was the percentage increase (to 2 s.f.) in the population during 1996?

   c) The projected population at the end of 2025 is 1.8×10^{10}. How many more people is this than at the end of 1995? (Give your answer in standard form to 2 s.f.)

8. The density of water is $1 \times 10^3 \text{ kg m}^{-3}$. Find (in standard form):

   a) the mass of water (in kg) in a cuboid measuring 2 m by 3 m by 5 m
   b) the volume (in m^3) of water whose mass is 5×10^8 tonnes (one tonne is 1000 kg)
   c) the volume (in cm^3) of $1 m^3$ of water
   d) the mass (in g) of $1 m^3$ of water
   e) the density of water (in $g \, cm^{-3}$)
   f) the mass of water (in g) in a cuboid measuring 6 cm by 3 cm by 10 cm.

9. The population of a certain country is 5.7×10^8 and its area is $7.21 \times 10^4 \text{ km}^2$. Find the population density (people per m^2) of this country in standard form (to 3 s.f.).

10. The diameter of the Earth is 1.3×10^7 m. Assuming that the Earth is a perfect sphere, find its circumference in km. Write the answer in standard form (to 2 s.f.).

11. The adult population of a country is 60 million. The average annual income per adult is $43,000. Find, in standard form, the total annual income from the adult population.

1.9 Rounding

YOU NEED TO:
- Round values to a specified degree of accuracy.
- Make estimates for calculations involving numbers, quantities and measurements.
- Round answers to a reasonable degree of accuracy in the context of a given problem.

◀ RECAP

When rounding to a given number of decimal places (d.p.), the rounded number must have exactly that number of digits after the decimal point.

13.565132   rounded to 1 d.p. is 13.6

                rounded to 2 d.p. is 13.57

When rounding to a given number of significant figures (s.f.), the first non-zero digit is the first significant figure.

13.565132   rounded to 1 s.f. is 10

                rounded to 5 s.f. is 13.565

0.004615   rounded to 1 s.f. is 0.005

                rounded to 3 s.f. is 0.00462

👁 WATCH OUT!
Remember that 5 rounds up.

👁 WATCH OUT!
With significant figures, remember to use zeros to keep the significant digits in their correct columns.

WORKED EXAMPLE

Find an approximate answer to $19.79 - 2.31 \times 3.15$ by rounding each number to 1 significant figure. **[2 marks]**

19.79 rounds to 20 (to 1 s.f.)

2.31 rounds to 2 (to 1 s.f.)

3.15 rounds to 3 (to 1 s.f.)

Hence the calculation becomes $20 - 2 \times 3 = 14$

🔑 KEY SKILLS
You need to be able to estimate the answer to a calculation by rounding.

👍 EXAM TIP
First, round each number to 1 s.f.

Don't forget the rules of BIDMAS when working out your final answer.

❓ QUESTIONS

1. **a)** Estimate the values of these expressions by rounding each number to 1 s.f.

    i) $\dfrac{4.1 + 3.9 \times 2.1}{3.2}$

    ii) $\dfrac{9.1 \times 8.1 + 3.8 \times 7.2}{10.1}$

   b) Calculate the values of the expressions in part **a)**, giving your answers to 3 s.f.

2. **a)** Find an estimate of the area of a circle of radius 9.97 cm.

   b) Calculate (to 3 s.f.) the area of a circle of radius 9.97 cm.

3. **a)** Without using your calculator, and showing all your working, estimate (to 1 s.f.) the answer to this calculation:

    $$\dfrac{2104.3 - (9.81)^2}{0.096}$$

   b) Using your calculator, find the answer to the calculation in part **a)** to 3 s.f.

1 Number

1.10 Upper and lower bounds

YOU NEED TO:

- Give upper and lower bounds for data rounded to a specified accuracy.
- **E** Find upper and lower bounds of the results of calculations that have used data rounded to a specified accuracy.

If the length of a rope is given as 5.3 m (to 2 s.f.), then you can calculate the lower and upper bounds for the length of the rope.

To calculate the lower and upper bounds, think of the two numbers (to 2 s.f.) immediately below and above 5.3. These are 5.2 and 5.4.

The lower bound is halfway between 5.2 and 5.3, so 5.25 is the lower bound.

The upper bound is halfway between 5.3 and 5.4, so 5.35 is the upper bound.

Bounds can be given using inequalities:

$5.25\,\text{m} \leq \text{length of rope} < 5.35\,\text{m}$

KEY SKILLS

You need to be able to give upper and lower bounds for rounded numbers.

WATCH OUT!

5.35 rounds up to 5.4 but you give the upper bound as 5.35 and use the 'less than' inequality. Don't give the upper bound as $5.34\dot{9}$ because $5.34\dot{9} = 5.35$. The answer 5.34 is incorrect.

WORKED EXAMPLE

Find lower and upper bounds for:

a) the perimeter of a circle given as 15 cm (to 2 s.f.)

b) the population of a city given as 340 000, correct to the nearest ten thousand. **[2 marks]**

a) $14.5\,\text{cm} \leq \text{perimeter} < 15.5\,\text{cm}$

b) $335\,000 \leq \text{population} < 345\,000$

EXAM TIP

Give your answers using inequalities.

Extended

WORKED EXAMPLE

The dimensions of a rectangle are 12 cm and 8 cm to the nearest cm. Calculate, to 3 s.f., the smallest possible area of the rectangle as a percentage of the largest possible area. **[4 marks]**

The lower and upper bounds for the 12 cm side are 11.5 cm and 12.5 cm, respectively.

The lower and upper bounds for the 8 cm side are 7.5 cm and 8.5 cm, respectively.

So the smallest possible area is $7.5 \times 11.5 = 86.25\,\text{cm}^2$.

The largest possible area is $8.5 \times 12.5 = 106.25\,\text{cm}^2$.

KEY SKILLS

You need to be able to calculate bounds for simple problems involving rounded values.

EXAM TIP

Write out the bounds of the numbers you are given. You will get a mark for this in the exam.

So the smallest possible area as a percentage of the largest possible area

$= \dfrac{86.25}{106.25} \times 100\%$

$= 81.2\%$ (to 3 s.f.)

RECAP

Use this table to ensure that you use the correct bounds in calculations.

Operation	Lower bound	Upper bound
Addition	LB + LB	UB + UB
Subtraction	LB − UB	UB − LB
Multiplication	LB × LB	UB × UB
Division	LB ÷ UB	UB ÷ LB

EXAM TIP

Notice that, for subtraction and division, what you are trying to do to the result is what you do to the **first** number in the calculation.

QUESTIONS

1. Mount Kenya is 17 060 ft high, correct to the nearest twenty feet. Find the smallest possible height of Mount Kenya.

2. Write down the upper and lower bounds for each of these numbers.
   a) $w = 73.43$ (to 2 d.p.)
   b) $x = 7320$ (to 3 s.f.)
   c) $y = 7320$ (to 4 s.f.)
   d) $z = 147.037$ (to 3 d.p.)
   e) $a = 100$ (to 3 s.f.)
   f) $b = 100$ (to 1 s.f.)

3. The distance between Nairobi and Dar es Salaam is 671 km.
   Find this distance:
   a) to the nearest 10 km
   b) to the nearest 20 km
   c) to the nearest 50 km.

4. The population of Nairobi in 2019 was estimated at 4 390 000 correct to the nearest ten thousand. Find the upper and lower bounds for the population.

Extended

5. A man runs a 100 m race and his time is measured as 10.3 s. If the track is accurate to the nearest metre and his time is accurate to the nearest 0.1 s, find the lower and upper bounds (to 1 d.p.) for his average speed.

6. The area of a rugby field is 6950 m², correct to 3 s.f. The length of the field is 95 m, correct to 2 s.f.
   a) Find the lower and upper bounds for the area of the field.
   b) Find the lower and upper bounds for the length of the field.
   c) Use the bounds from parts **a)** and **b)** to calculate the lower and upper bounds (to 3 s.f.) for the width of the field.

7. The formula for the distance s travelled by a body with initial speed u and constant acceleration a after a time t is given by $s = ut + \dfrac{1}{2}at^2$. Find the least and greatest possible values (to 3 s.f.) of s when $u = 6.1$, $a = 4.5$ and $t = 13.6$, all correct to 1 d.p.

8. Pythagoras' theorem states that $a^2 + b^2 = c^2$ where a, b and c are the lengths of the three sides of a right-angled triangle and c is the hypotenuse.
   If $a = 4.3$ cm and $c = 12.1$ cm, both correct to 1 d.p, find the smallest and largest possible values for b (to 1 d.p.).

9. The formula $s = \dfrac{v^2 - u^2}{2a}$ is used to find the distance travelled by an object whose initial speed is u, whose final speed is v, and whose acceleration is a. Find an inequality for s (to 2 s.f.) if $v = 15$, $u = 11$ and $a = 2.3$, all correct to 2 s.f.

To **Raise your grade** now try question 2 on page 50

1.11 Ratio and proportion

> **YOU NEED TO:**
> - Understand and use ratio and proportion to:
>   give ratios in their simplest form, divide a quantity in a given ratio, use proportional reasoning and ratios in context.

WORKED EXAMPLE

a) Simplify the ratio $150:100$
b) Divide 64 into the ratio $5:4:7$ **[3 marks]**

a) $150:100$
$= 3:2$

b) $5 + 4 + 7 = 16$
$64 \div 16 = 4$
$5 \times 4 = 20$
$4 \times 4 = 16$
$7 \times 4 = 28$

Hence 64 divided into the ratio $5:4:7$ is $20:16:28$

WORKED EXAMPLE

Peter buys 3 kg of carrots from a market and pays $4.50.

Lotte buys 5 kg of carrots from the same market.

Calculate the cost of Lotte's carrots. **[2 marks]**

3 kg costs $4.50

1 kg costs $1.50

5 kg costs $1.50 \times 5 = $7.50

🔑 KEY SKILLS

You need to be able to simplify ratios and divide amounts in a given ratio.

👍 EXAM TIP

For part **a)**, divide both numbers by their highest common factor to simplify the ratio as much as possible.

For part **b)**, add up the different parts of the ratio and divide the total by the result to find the value of each part. Then multiply the value of each part by the number of parts.

👍 EXAM TIP

Work out the cost of 1 kg of carrots.

Multiply the **unit cost** by 5 to find the cost of Lotte's carrots.

🔑 KEY SKILLS

You need to be able to use the **unitary method** to solve simple problems involving proportion.

1.11 Ratio and proportion

WORKED EXAMPLE

The supermarket sells 1.5 kg of apples for $0.60.

The local shop sells 1.75 kg of apples for $0.75.

Which is the better value? [3 marks]

Supermarket: $0.60 ÷ 1.5 = $0.40 per kg

Local shop: $0.75 ÷ 1.75 = $0.43 per kg (to 2 d.p.)

The supermarket is better value.

EXAM TIP

You could divide the number of kilograms by the price, but then you would get 'kilograms per dollar' and the higher value would then be the better value.

Make sure you show that you know how to interpret your calculations.

WORKED EXAMPLE

a) Increase $30 in the ratio 5 : 4
b) Decrease 54 kg in the ratio 5 : 9 [2 marks]

a) $\frac{5}{4} = \frac{x}{30}$

$\Rightarrow x = 30 \times \frac{5}{4} = \$37.50$

b) $\frac{5}{9} = \frac{x}{54}$ kg

$\Rightarrow x = 54 \times \frac{5}{9} = 30$ kg

KEY SKILLS

You need to be able to increase and decrease a quantity in a given ratio.

EXAM TIP

Write the ratio as a fraction equal to the new amount divided by the original amount.

The approach works whether increasing or decreasing by a given ratio.

QUESTIONS

1. Anatole, Brij and Christophe receive $560 from their great aunt, to be divided in the ratio Anatole : Brij : Christophe = 3 : 5 : 6.
   a) Calculate how much each receives.
   b) Christophe puts all his share into a venture with Dimitri. If Dimitri adds to his share and puts in $300 altogether, find the ratio of Christophe's investment to Dimitri's investment.

2. When David's car was repaired, the charge for labour was $200.

   This was $\frac{4}{7}$ of the total bill. What was the total bill?

3. The local supermarket sells 6 eggs for $2.25. The local farm shop sells 10 eggs for $3.50. Which is the better value?

4. a) Increase $50 in the ratio 7 : 5.
   b) Decrease 45 grams in the ratio 8 : 9.

To **Raise your grade** now try question 5 on page 50

37

1 Number

1.12 Rates of change

YOU NEED TO:
- Use common measures of rate.
- Apply other measures of rate.
- Solve problems involving average speed.

◀◀ RECAP

average speed = $\dfrac{\text{total distance travelled}}{\text{total time taken}}$

or $S = \dfrac{D}{T}$

This can be rearranged to make either D or T the subject:

$D = S \times T$ or $T = \dfrac{D}{S}$

WORKED EXAMPLE

A car travels 189 km at an average speed of 60 km h^{-1}.

How long does the journey take? **[2 marks]**

$T = \dfrac{D}{S} = \dfrac{189}{60}$

$\dfrac{189}{60} = 3.15$ hours

3.15 hours = 3 hours and 9 minutes

WORKED EXAMPLE

A 90 litre tank is filled full of water in 2 minutes and 15 seconds.

What is the rate of flow (in litres per minute) of the water into the tank? **[2 marks]**

2 minutes and 15 seconds = 2.25 minutes

$\dfrac{90}{2.25} = 40$ litres per minute

❓ QUESTIONS

1. A coach leaves London at 06:55 and arrives in Glasgow at 16:12, a distance of 667 km. Find the average speed in kilometres per hour.

2. A plane travels from Windhoek to Johannesburg in 1 h 45 min. If the distance is 1190 km, find the average speed.

3. A bucket is filled with water at a rate of 0.04 litres per second. If it takes 9 minutes to fill the bucket, what is the capacity of the bucket?

To **Raise your grade** now try questions 6 and 7 on page 50

👁 WATCH OUT!

Make sure you work in consistent units. For example, if speed is in m s^{-1}, make sure distance is in metres and time in seconds.

👍 EXAM TIP

When dealing with compound units, the units tell you the formula. For example, speed is measured in 'metres per second', which is 'distance divided by time'.

Use the formula for time in terms of D and S.

Work your answer out as a decimal number of hours and then convert to hours and minutes.

You can use the hours and minutes button on your calculator to do this.

👁 WATCH OUT!

Remember that there are 60, not 100, minutes in an hour.

👍 EXAM TIP

Look at the rate required in the question. In this case, it is litres per minute, which means you need to divide the number of litres by the number of minutes.

🔑 KEY SKILLS

You need to be able to work with other measures of rate.

1.13 Percentages

YOU NEED TO:
- Calculate a given percentage of a quantity.
- Express one quantity as a percentage of another.
- Calculate percentage increase or decrease.
- Calculate with simple and compound interest.
- **E** Calculate using reverse percentages.

◀◀ RECAP
A percentage multiplier can be used to find a percentage of a given quantity.

For example, to find 67% of 200 kg, multiply 200 by 0.67

WORKED EXAMPLE
A road is 1300 metres long and 12% of it requires resurfacing.

Calculate the length of road that needs resurfacing. **[2 marks]**

$1300 \times 0.12 = 156$ metres

◀◀ RECAP
To find one quantity as a percentage of another, you divide the first quantity by the second and then multiply by 100%.

WORKED EXAMPLE
A student scored 75 out of 120 in a test.

Express this as a percentage. **[2 marks]**

Percentage $= \dfrac{75}{120} \times 100\%$

$= 62.5\%$

◀◀ RECAP
To calculate a percentage increase or decrease, use a percentage multiplier.

For example, to find an increase of 23%, the multiplier is $(1 + 0.23) = 1.23$.

To find a decrease of 25%, the multiplier is $(1 - 0.25) = 0.75$.

✏️ APPLY
Write down the multiplier you would use to find:
a) 25% of 120
b) 40% of 300
c) 82% of 450
d) 91% of 1250
e) 125% of 560

👍 EXAM TIP
Make sure you write down the calculation you are going to do, even if you use a calculator.

👍 EXAM TIP
You are asked to find 75 as a percentage of 120 so divide 75 by 120 and then multiply by 100%.

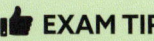

1 Number

WORKED EXAMPLE

a) One year, a football club's average crowd attendance was 41 200. The following year the attendance rose by 4%.
What was its new average attendance?

b) A car worth $5000 loses 15% of its value in a year.
What is it worth after one year? **[4 marks]**

a) $41200 \times 1.04 = 42\,848$

b) $5000 \times 0.85 = \$4250$

> 👍 **EXAM TIP**
>
> A percentage **increase** of 4% converts to a multiplier of $1 + 0.04 = 1.04$.
>
> A percentage **decrease** of 15% converts to a multiplier of $1 - 0.15 = 0.85$.

Extended

WORKED EXAMPLE

a) The price of a skirt is reduced by 10% in a sale.
If it cost $31.50 in the sale, what was the original price of the skirt?

b) The price of a camera including sales tax of 17.5% is $94.
What was the price before sales tax was added? **[4 marks]**

a) Let the original cost be x.
$x \times 0.9 = 31.50$
$x = \dfrac{31.5}{0.9} = 35$
so the original price of the skirt was $35.

b) Let the pre-tax cost of the camera be y.
$y \times 1.175 = 94$
$y = \dfrac{94}{1.175} = 80$
so the original price of the camera before sales tax was added was $80.

> 🔑 **KEY SKILLS**
>
> You need to be able to find the original quantity when you are given the quantity after a percentage increase or decrease. These are known as **reverse percentages**.

> 👍 **EXAM TIP**
>
> Set up and solve an equation with the original price, x, and the multiplier for a 10% decrease, 0.9.
>
> You can use the equation method for both a percentage increase and for a percentage decrease.

If you have a savings account or an investment, you usually get interest paid to you on these.

If you have a loan or a credit card, you usually pay interest on what you borrow.

The amount you invest or borrow is called the **principal**.

Interest falls into two categories:

Simple interest is paid only on the original principal.

Compound interest is added to the principal at the end of each period of time (usually a year) and the next year's interest is paid on the new amount.

1.13 Percentages

WORKED EXAMPLE

Joseph invests $500 at 4% per annum simple interest.

How much money will he have at the end of 5 years? **[2 marks]**

Interest = 4% of $500 = 0.04 × 500 = $20 per year

Total interest in 5 years = $20 × 5 = $100

Joseph will have $500 + $100 = $600

👍 EXAM TIP

'Per annum' means each year.

Simple interest is paid only on the principal, so it will be the same amount each year.

WORKED EXAMPLE

Ali invests $200 at 3% compound interest. How much money will he have after 3 years? **[2 marks]**

3% of $200 = $6

$200 + $6 = $206, so Ali has $206 at the start of year 2.

3% of $206 = $6.18

$206 + $6.18 = $212.18, so Ali has $212.18 at the start of year 3.

3% of $212.18 = $6.37 to 2 d.p.

$212.18 + $6.37 = $218.55

Ali will have $218.55 after 3 years.

👍 EXAM TIP

The interest is added to the principal each year.

You can simplify this working by calculating 200×1.03^3.

⏪ RECAP

The formula for compound interest is:

Amount after n years = $P \times$ (percentage multiplier)^n

where P is the principal and n is the number of years.

WORKED EXAMPLE

Use the formula for compound interest to calculate:

a) how much Rose owes on a bank loan of $800 after 2 years if the rate of interest she pays is 7%

b) how much an $18 000 car is worth after 3 years if its value depreciates by 30% per year. **[4 marks]**

a) $800 \times 1.07^2 = 915.92$

Rose owes $915.92 after two years.

b) $18000 \times 0.7^3 = 6174$

The car is worth $6174 after 3 years.

1 Number

❓ QUESTIONS

1. A man buys plane tickets for himself, his wife and his four children.

   The adult fare is $172 and the child fare is 67% of the adult fare.

   Find the total cost of the journey.

2. In April, a lawnmower cost $265. In the September sale, it was only $225.25.

   What was the percentage discount?

3. A bleach bottle is labelled '900 ml for the price of 750 ml: x% extra free'.

   The x is smudged and illegible. What is x?

4. The profit of a company in 2004 was $1 500 000. In 2005 the profit was 25% higher than it was in 2004 but in 2006 the profit fell by 40%.
   a) Show that the profit in 2005 was $1 875 000.
   b) What was the profit in 2006?

5. A bank offers 5% compound interest on investments. A person invests £2000.
   a) What is his investment worth after 2 years?
   b) What is the total percentage increase?

6. An investment fund has increased in value by a total of 21% over the last two years.
   a) A person invested £1000 in the fund two years ago. What is it worth now?
   b) Calculate the yearly rate of interest assuming that it was:
      i) compound interest
      ii) simple interest.

7. A bank offers 2% simple interest per year. A person opens an account with a deposit of €750. They close the account 11 months later. How much money do they withdraw?

8. Jack invests €80 in an account offering him 3.6% simple interest. He removes his money after 10 months. How much interest does he get?

9. The sale price of a garden table is $48 and it has a sign saying 'Reduction of 20%'.

   What was the price of the table before the sale?

10. A school claims that the students' average marks in an exam have increased by 15% over 5 years. Two boys are told that the average mark is now 85.1. George thinks that the average mark five years ago was 72.335 but James thinks it was 74. Who is correct, and how is the correct answer obtained?

11. In a sale, all items are reduced by 15%. A carpet now costs $15.30 per square metre. What was the price before the sale?

12. A document is photocopied so that the lengths of the copy are 70% of the original lengths. If the copy measures 12.6 cm by 17.5 cm, what are the dimensions of the original document?

13. The number of students at a school in 2024 was 85% of the number at the school in 2023. In 2024, the number of students was 1020. How many students were there in 2023?

14. Find the original price of a car which was sold for $1200 at a loss of 4%.

15. Find the original price of an antique which was sold at $545 at a profit of 9%.

> To **Raise your grade** now try question 4 on page 50

1.14 Using a calculator

YOU NEED TO:
- Use a calculator efficiently.
- Enter values appropriately on a calculator.
- Interpret the calculator display appropriately.

WORKED EXAMPLE

a) Estimate the answer to $\sqrt{7.62 + 8.13}$ by first rounding each number to 1 s.f.

b) Calculate the exact answer to $\sqrt{7.62 + 8.13}$, giving it as a decimal, and showing all of the digits on your calculator display. **[3 marks]**

a) $7.62 = 8$ (to 1 s.f.) and $8.13 = 8$ (to 1 s.f.)
$\Rightarrow \sqrt{7.62 + 8.13} \approx \sqrt{8 + 8}$
$= \sqrt{16}$
$= 4$

b) $\sqrt{7.62 + 8.13} = 3.968626967$

KEY SKILLS
You need to be able to use your calculator correctly and efficiently to carry out complex calculations.

You also need to be able to check your answer to a calculation using a suitable estimate.

EXAM TIP
Write down your approximations and show all steps in the working.

QUESTIONS

1. a) Estimate the values of these expressions by rounding all the numbers to 1 s.f.

   i) $\dfrac{14.1 + 6.9 \times 8.3}{8.9 - 2.1}$

   ii) $\dfrac{10.1^2 + 2.1 \times 20.7}{23.1 - 3.2}$

   b) Calculate the values of the expressions in part **a)**, giving your answers to 3 s.f.

2. Calculate the value of each expression (to 3 s.f.).

   a) $\sqrt{7.2 \times 3.5 + 2.1 \times 5.7}$

   b) $\dfrac{6.1 - 3.5}{2.1 + 4.7 \times 1.8}$

   c) $\dfrac{2.7^2 + 4.2^2}{1.9^4}$

   Check your answers using sensible estimates.

WATCH OUT!
There are many different calculators available for use in your exam. Make sure that you can use the one that you have rather than following a set series of instructions or key strokes from a textbook.

1 Number

1.15 Calculating with time

YOU NEED TO:
- Calculate with time: seconds (s), minutes (min), hours (h), days, weeks, months, years, including the relationship between units.
- Calculate times in terms of the 24-hour and 12-hour clock.
- Read clocks and timetables.

WORKED EXAMPLE

a) A flight departed at 15:14 and arrived at 18:42. How long was the flight?

b) A girl arrived at a rehearsal at 19:35 and left at 21:29.
   For how long was she at the rehearsal? **[2 marks]**

a) Three hours on from 15:14 is 18:14.
   18:42 is another 28 minutes on from that.
   So the flight took 3 h 28 min.

b) One hour on from 19:35 is 20:35.
   20:35 to 21:00 is another 25 minutes. 21:00 to 21:29 is a further 29 minutes.
   25 + 29 = 54
   So she was at the rehearsal for 1 hour 54 minutes.

KEY SKILLS
You need to be able to calculate with time.

EXAM TIP
Break the flight time down into whole hours, and then minutes.

EXAM TIP
Timetables may be given using either the 24-hour clock or the 12-hour clock.

WORKED EXAMPLE

Derek wanted to get from Twickenham to Wembley Stadium.
A website gave him the information shown in the table.

a) How long did the journey take in total?
b) At which station did Derek wait for 8 minutes? **[2 marks]**

a) He departed at 11:58 and arrived at 13:02. One hour after 11:58 is to 12:58 so add another 4 minutes.
   So the journey took 1 hour 4 minutes.
b) There is an 8 minute gap between 12:03 and 12:11.
   So he waited for 8 minutes at Richmond Rail Station.

Instruction	Time
Depart Twickenham	11:58
Arrive at Richmond	12:03
Depart Richmond	12:11
Arrive at Willesden Junction	12:30
Depart Willesden Junction	12:35
Arrive at Wembley Central	12:42
Depart Wembley Central	12:48
Arrive at Wembley Stadium	13:02

QUESTIONS

1. A plane leaves Nairobi airport at 23:30 and arrives in London the next day at 05:20. The time in Nairobi is 3 hours ahead of the time in London.
   a) How long does the flight take?

   The return flight leaves London at 10:05 and arrives in Nairobi at 21:35.

   b) How long does the return flight take?
   c) Calculate how much longer the outward journey is than the return journey.

2. A flight from Singapore to London leaves at 01:30 local time and arrives the same day at 05:55 local time. The airline website says that journey takes 12 h 25 min.
   a) How many hours ahead of London is Singapore?
   b) A traveller arriving at Singapore phones home when the time in Singapore is 07:00.
   What is the time in London when he phones?

1.16 Calculating with money

YOU NEED TO:
- Calculate with money.
- Convert from one currency to another.

When you travel abroad, you often have to exchange your domestic currency for the currency of the country to which you are travelling.

You do this using an **exchange rate**.

Exchange rates change frequently and they may also depend on where you change your currency, for example, at the airport or using a local bank.

WORKED EXAMPLE

Susan changed some Canadian dollars into Kenyan shillings. The exchange rate was 0.01592 Canadian dollars to one Kenyan shilling.

a) How many Kenyan shillings did Susan get if she exchanged 500 Canadian dollars?

b) At the end of her trip, she exchanged 2000 Kenyan shillings back into Canadian dollars. The exchange rate was the same. How many Canadian dollars did she get? **[4 marks]**

a) Susan got one Kenyan shilling for every 0.01592 Canadian dollars.
So she got $\frac{500}{0.01592}$ = 31 407 Kenyan shillings (to the nearest shilling)

b) For each Kenyan shilling she got 0.01592 Canadian dollars.
So she got 2000 × 0.01592 = 31.84 Canadian dollars.

🔑 KEY SKILLS
You need to be able to calculate currency conversions.

✏️ APPLY
At the time of writing this book, the exchange rate from GBP (£) to USD ($) was £1 to $1.28.

Investigate the exchange rate of your local currency.

Find at least five different foreign currencies to compare.

👍 EXAM TIP
You will be told the exchange rate to use in an exam question.

❓ QUESTIONS

1. A Bureau de Change offers $1.568 per £ sterling but charges a commission fee of £3.

   How many dollars (to the nearest cent) does Joe get for £75?

2. If £1 = €1.27 then find:
   a) the cost (to the nearest £) of a holiday house which costs €450 per week to rent
   b) how much (in €) a British holidaymaker would get for £300 at the foreign exchange.

3. Mr Smith converts 600 euros into Chinese yuan. The bank offers an exchange rate of 1 euro for 8.37 yuan.
   a) How many yuan does he get?
   b) At the end of his holiday Mr Smith has 100 Chinese yuan. If he changes his yuan back into euros at the same rate of exchange, how many euros does he get?

1.17 Exponential growth and decay

YOU NEED TO:

- **E** Use exponential growth and decay.

Extended

The principle of compound interest can be extended to model growth and decay in many different situations. For example, the growth or decline of a population of organisms, the decay of a radioactive element and the value of a financial product such as shares.

◀◀ RECAP

The formula for **exponential growth** (or **decay**) is:

amount after n time intervals = $A \times$ (growth or decay factor)^n

where A is the initial amount and the growth or decay factor is a percentage multiplier.

KEY SKILLS

You need to be able to calculate exponential growth and decay.

WORKED EXAMPLE

A colony of bacteria increases at a rate of 10% per hour.

Initially there are 120 bacteria.

Calculate the number of bacteria in the colony after 8 hours. **[2 marks]**

Number of bacteria after 8 hours = $120 \times 1.1^8 = 257$

EXAM TIP

Use the formula for exponential growth with a growth factor of $1 + 0.1 = 1.1$.

Your answer must be a whole number since you are dealing with organisms.

WORKED EXAMPLE

A radioactive substance decays at a rate of 7% per day.

Initially there are 1000 atoms of the substance.

Calculate the number of atoms after 5 days. **[2 marks]**

Number of atoms after 5 days = $1000 \times 0.93^5 = 696$.

EXAM TIP

The multiplier represents a percentage decrease since you are modelling decay.

The answer should again be a whole number.

1.17 Exponential growth and decay

WORKED EXAMPLE

The value of a financial bond increases at a rate of 4% for the first 6 months, and then decreases at a rate of 2% for the next 4 months.

The bond is initially worth $500.

Calculate the value of the bond after 10 months. **[3 marks]**

Value of bond after 10 months
$= 500 \times 1.04^6 \times 0.98^4 = \$583.54$.

EXAM TIP

You can combine the growth and the decay parts into a single calculation using the two different multipliers.

Make sure that you show all your working, even if you are using a calculator.

QUESTIONS

1. The population of a colony of rabbits grows at a rate of 20% per month.

   Initially there are 70 rabbits.

   Calculate the number of rabbits after 7 months.

2. A colony of bacteria is treated with an antibiotic that reduces the number of bacteria by 30% every hour. Initially, there are 3000 bacteria.

   a) Calculate the number of bacteria after 3 hours.

   b) After how many hours will the bacteria colony be less than 200?

3. A radioactive substance decays at a rate of 14% per year. Initially, there are 2000 atoms of the substance.

   a) Calculate the number of atoms after 2 years.

   The 'half-life' of a radioactive substance is the time it takes for half of the atoms to decay.

   b) Calculate the half-life for this radioactive substance. Give your answer as a decimal number of years, correct to 1 d.p.

4. A financial product decreases in value by 7% each month for the first 3 months, and then increases in value by 7% for the next 3 months.

   Adil says that the product is now worth the same as it was at the start of the 6-month period.

   a) Show that Adil is wrong.

   b) Find, correct to 2 d.p., the percentage change in the value of the product over the 6-month period.

1.18 Surds

> **YOU NEED TO:**
> - **E** Understand and use surds, including simplifying expressions.
> - **E** Rationalise the denominator.

Extended

◀◀ RECAP

Surds are irrational roots of numbers, or expressions that contain these, such as:

$\sqrt{2}, 3 + \sqrt{5}, \dfrac{1 + \sqrt{7}}{4}$

WORKED EXAMPLE

Simplify the following surds.

a) $\sqrt{12}$ **[1 mark]**

b) $\sqrt{50} - \sqrt{8}$ **[2 marks]**

a) $\sqrt{12} = \sqrt{4} \times \sqrt{3} = 2\sqrt{3}$

b) $\sqrt{50} - \sqrt{8} = 5\sqrt{2} - 2\sqrt{2} = 3\sqrt{2}$

👁 WATCH OUT!

Remember that $\sqrt{a} \times \sqrt{b} = \sqrt{ab}$ and $\sqrt{a} \div \sqrt{b} = \sqrt{a \div b}$
but $\sqrt{a} + \sqrt{b} \neq \sqrt{a+b}$ and $\sqrt{a} - \sqrt{b} \neq \sqrt{a-b}$

👁 WATCH OUT!

Don't forget to simplify your answer as much as possible, particularly if the question asks you to!

🔑 KEY SKILLS

You need to be able to simplify surds.

👍 EXAM TIP

Look for square factors of the surd you are trying to simplify.

For part **b)**, surds of the same type can be collected together as you would collect variables when doing algebra.

1.18 Surds

WORKED EXAMPLE

Rationalise the denominator:

a) $\dfrac{7}{\sqrt{2}}$ [2 marks]

b) $\dfrac{5}{2\sqrt{3}}$ [2 marks]

c) $\dfrac{6}{3-\sqrt{2}}$ [2 marks]

a) $\dfrac{7}{\sqrt{2}} \times \dfrac{\sqrt{2}}{\sqrt{2}} = \dfrac{7\sqrt{2}}{2}$

b) $\dfrac{9}{2\sqrt{3}} \times \dfrac{\sqrt{3}}{\sqrt{3}} = \dfrac{9\sqrt{3}}{6} = \dfrac{3\sqrt{3}}{2}$

c) $\dfrac{6}{3-\sqrt{2}} \times \dfrac{3+\sqrt{2}}{3+\sqrt{2}} = \dfrac{18+6\sqrt{2}}{9-2} = \dfrac{18+6\sqrt{2}}{7}$

KEY SKILLS
You need to be able to rationalise the denominator of a surd.

EXAM TIP
In **b)**, only multiply top and bottom by the irrational part of the denominator, not by the whole denominator.

EXAM TIP
The idea being used in part **c)** is the difference of two squares. This is just one of its many applications.

QUESTIONS

1. Simplify:
   a) $\sqrt{20}$
   b) $\sqrt{128}$
   c) $\sqrt{108}$
   d) $\sqrt{175}$
   e) $\sqrt{32}$
   f) $\sqrt{1400}$

2. Simplify:
   a) $\sqrt{12} + \sqrt{48}$
   b) $\sqrt{125} - \sqrt{80}$
   c) $10\sqrt{2} - \sqrt{200}$

3. Rationalise the denominator:
   a) $\dfrac{3}{\sqrt{5}}$
   b) $\dfrac{4}{\sqrt{11}}$
   c) $\dfrac{7}{\sqrt{10}}$
   d) $\dfrac{8}{5\sqrt{2}}$
   e) $\dfrac{12}{5\sqrt{3}}$
   f) $\dfrac{21}{2\sqrt{7}}$

4. Rationalise the denominator:
   a) $\dfrac{4}{1+\sqrt{3}}$
   b) $\dfrac{2}{\sqrt{3}+5}$
   c) $\dfrac{8}{\sqrt{7}-\sqrt{6}}$
   d) $\dfrac{3+\sqrt{2}}{1+\sqrt{2}}$
   e) $\dfrac{\sqrt{7}-1}{\sqrt{7}-2}$
   f) $\dfrac{4+\sqrt{5}}{2-\sqrt{5}}$

1 Number

↑ Raise your grade

1. Work out the value of $3 + \dfrac{1}{7 + \dfrac{1}{16}}$. Give your answer correct to 6 decimal places. **[2 marks]**

2. A man walked 1000 miles, correct to the nearest 10 miles. If he walked on average 10 miles per day, correct to the nearest mile, what are the lower and upper bounds, to the nearest day, for the number of days he spent walking? **[4 marks]**

3. How many prime numbers are there between 90 and 100? **[1 mark]**

4. In 2016, Pete bought a car for $20 000.
   a) This was 25% more than he paid for his previous car. How much did he pay for his previous car? **[2 marks]**
   b) One year later, Pete's new car was worth the same as he paid for his old car. By what percentage had his new car decreased in value? **[2 marks]**

5. Genoveva and Apolonia share a large container of cherries in the ratio 3 : 5. If Apolonia gives Genoveva 20 cherries from her share, their new shares are in the ratio 7 : 5. How many cherries are there altogether? **[3 marks]**

6. The Titanic ocean liner was 2.7×10^2 m long. If a domestic cat can run at a speed of $40 \, \text{km h}^{-1}$, how long would it have taken a cat to run from one end of the Titanic to the other? Give your answer to the nearest second. **[2 marks]**

7. If a train travels for 50 km at a speed of $40 \, \text{km h}^{-1}$, then travels a further 60 km at a speed of $30 \, \text{km h}^{-1}$, find:
   a) the total time taken **[3 marks]**
   b) the average speed at which the train travelled, correct to 3 significant figures. **[3 marks]**

2 Algebra and graphs

Your revision checklist

Tick these circles to build a record of your revision.

Core/**E** Extended syllabus   ☹ 😐 🙂

2.1	Know that letters can be used to represent generalised numbers.	○ ○ ○
	Substitute numbers into expressions and formulas.	○ ○ ○
2.2	Simplify expressions by collecting like terms.	○ ○ ○
	Expand products of algebraic expressions.	○ ○ ○
	Factorise by extracting common factors.	○ ○ ○
	E Factorise expressions of the form: $ax + bx + kay + kby$, $a^2x^2 - b^2y^2$, $a^2 + 2ab + b^2$, $ax^2 + bx + c$, $ax^3 + bx^2 + cx$.	○ ○ ○
	E Complete the square for expressions in the form $ax^2 + bx + c$.	○ ○ ○
2.3	**E** Manipulate algebraic fractions.	○ ○ ○
	E Factorise and simplify rational expressions.	○ ○ ○
2.4	Understand and use indices (positive, zero and negative).	○ ○ ○
	Understand and use the rules of indices.	○ ○ ○
	E Understand and use fractional indices.	○ ○ ○
2.5	Construct expressions, equations and formulas.	○ ○ ○
	Solve linear equations in one unknown.	○ ○ ○
	E Solve fractional equations with numerical and linear algebraic denominators.	○ ○ ○
	Solve simultaneous linear equations in two unknowns.	○ ○ ○
	E Solve simultaneous equations, involving one linear and one non-linear.	○ ○ ○
	E Solve quadratic equations by factorisation, completing the square and by use of the quadratic formula.	○ ○ ○
	Change the subject of formulas.	○ ○ ○
2.6	Represent and interpret inequalities, including on a number line.	○ ○ ○
	E Construct, solve and interpret linear inequalities.	○ ○ ○
	E Represent and interpret linear inequalities in two variables graphically.	○ ○ ○
	E List inequalities that define a given region.	○ ○ ○
2.7	Continue a given number sequence or pattern.	○ ○ ○
	Recognise patterns in sequences, including the term-to-term rule, and relationships between different sequences.	○ ○ ○
	Find and use the nth term of linear, simple quadratic and simple cubic sequences.	○ ○ ○
	E Find and use the nth term of exponential sequences.	○ ○ ○
2.8	**E** Express direct and inverse proportion in algebraic terms and use this form of expression to find unknown quantities.	○ ○ ○
2.9	Use and interpret graphs in practical situations including travel graphs and conversion graphs.	○ ○ ○
	Draw graphs from given data.	○ ○ ○
	E Apply the idea of rate of change to simple kinematics involving distance–time and speed–time graphs, acceleration and deceleration.	○ ○ ○
	E Calculate distance travelled as area under a speed–time graph.	○ ○ ○
2.10	Construct tables of values, and draw, recognise and interpret graphs for functions of the following forms: $ax + b$, $\pm x^2 + ax + b$, $\dfrac{a}{x}(x \neq 0)$, where a and b are integer coefficients.	○ ○ ○
	E Construct tables of values, and draw, recognise and interpret graphs for functions of ax^n (includes sums of no more than three of these) and $ab^x + c$, where $n = -2, -1, -\dfrac{1}{2}, 0, \dfrac{1}{2}, 1, 2, 3$; a and c are rational numbers; and b is a positive integer.	○ ○ ○
	Solve associated equations graphically, including finding and interpreting roots by graphical methods.	○ ○ ○
	E Draw and interpret graphs representing exponential growth and decay problems.	○ ○ ○
2.11	Recognise, sketch and interpret graphs of linear and quadratic functions.	○ ○ ○
	E Recognise, sketch and interpret graphs of cubic, reciprocal and exponential functions.	○ ○ ○
2.12	**E** Estimate gradients of curves by drawing tangents.	○ ○ ○
	E Use the derivatives of functions of the form ax^n, where a is a rational constant and n is a positive integer or zero, and simple sums of not more than three of these.	○ ○ ○
	E Apply differentiation to gradients and stationary points (turning points).	○ ○ ○
	E Discriminate between maxima and minima by any method.	○ ○ ○
2.13	**E** Understand functions, domain and range and use function notation.	○ ○ ○
	E Understand and find inverse functions $f^{-1}(x)$.	○ ○ ○
	E Form composite functions as defined by $gf(x) = g(f(x))$.	○ ○ ○

2 Algebra and graphs

2.1 Introduction to algebra

YOU NEED TO:
- Know that letters can be used to represent generalised numbers.
- Substitute numbers into expressions and formulas.

◀◀ RECAP

In algebra, you use letters to represent numbers, either because you don't know their value or because you want to be able to change their value.

WORKED EXAMPLE

If $D = \dfrac{(a-b)^2}{c^3}$, find the value of D when $a = 7$, $b = -2$ and $c = 3$. **[2 marks]**

$$D = \dfrac{(7-(-2))^2}{3^3}$$
$$= \dfrac{9^2}{3^3}$$
$$= 3$$

👍 KEY SKILLS

You must be able to substitute numbers for words and letters in formulas.

👍 EXAM TIP

Remember to use the correct order of operations when solving questions like this one.

👍 EXAM TIP

Here are some general guidelines for rearranging formulas.

Step 1 Clear the fractions.
Step 2 Multiply out any brackets involving x.
Step 3 Put all x-terms together on one side of the equation.
Step 4 Divide both sides by the coefficient of the x-term.

WORKED EXAMPLE

Jane thinks of a number, multiplies it by 4, adds 7 and gets the answer 19.

Construct an equation to represent this situation and then solve it. **[3 marks]**

Let x represent Jane's number.

$4x + 7 = 19$

Then

$4x = 19 - 7 = 12$

$x = \dfrac{12}{4} = 3$

Therefore Jane's number was 3.

🔑 KEY SKILLS

You must be able to construct simple expressions and set up simple equations.

QUESTIONS

1. The acceleration of a body, moving with uniform acceleration, is given by the formula
$$a = \frac{v - u}{t}.$$
In this formula, a is the uniform acceleration, u is the initial velocity and v is the velocity at time t.

   Use the formula to find a (in m s^{-2}) if:

   a) $v = 5\,\text{m s}^{-1}$, $u = 1\,\text{m s}^{-1}$, $t = 8\,\text{s}$

   b) $v = 0.1\,\text{m s}^{-1}$, $u = 0.02\,\text{m s}^{-1}$, $t = 2\,\text{s}$

2. Use the formula $s = vt - \frac{1}{2}at^2$ to calculate s (to 2 s.f.) given that $v = 27.27\,\text{m s}^{-1}$, $a = 9.81\,\text{m s}^{-2}$ and $t = 1.73\,\text{s}$.

3. Evaluate $a^2 b$ when $a = 5000$ and $b = 300$.

4. Given that $s = \dfrac{v^2 - u^2}{2a}$, find s when $u = 60$, $v = 80$ and $a = 400$.

5. If $\sqrt{\dfrac{py + q}{r}} = s$, find y (to 3 s.f.) when $p = 132$, $q = 251$, $r = 158$ and $s = 17$.

To **Raise your grade** now try question 3 on page 95

2 Algebra and graphs

2.2 Manipulating algebraic expressions

YOU NEED TO:

- Simplify expressions by collecting like terms.
- Expand products of algebraic expressions.
- Factorise by extracting common factors.
- **E** Factorise expressions of the form: $ax + bx + kay + kby$, $a^2x^2 - b^2y^2$, $a^2 + 2ab + b^2$, $ax^2 + bx + c$, $ax^3 + bx^2 + cx$.
- **E** Complete the square for expressions in the form $ax^2 + bx + c$.

WORKED EXAMPLE

Expand these expressions.

a) $4a(2a + 3b)$ [1 mark]
b) $5p(2p - 5q)$ [1 mark]
c) $6s(2s + 7t) - 2s(5s - 2t)$ [1 mark]

a) $4a(2a + 3b) = 8a^2 + 12ab$ ($8a^2$ is shorthand for $8 \times a \times a$.)
b) $5p(2p - 5q) = 10p^2 - 25pq$
c) $6s(2s + 7t) - 2s(5s - 2t)$
   $= 12s^2 + 42st - 10s^2 + 4st$
   $= 2s^2 + 46st$

KEY SKILLS

You must be able to expand products of algebraic expressions and reverse the process by factorising.

EXAM TIP

Remember: you can only collect together terms that contain the same variables raised to the same powers.

WORKED EXAMPLE

Expand these expressions.

a) $(x + 2)(x + 5)$
b) $(x + 7)(x - 4)$
c) $(x - 3)(x - 5)$
d) $(2x + 3)(5x - 9)$ [4 marks]

Use the FOIL method:

a) $(x + 2)(x + 5) = x \times x + x \times 5 + 2 \times x + 2 \times 5$
   $= x^2 + 5x + 2x + 10$
   $= x^2 + 7x + 10$

b) $(x + 7)(x - 4) = x \times x + x \times (-4) + 7 \times x + 7 \times (-4)$
   $= x^2 - 4x + 7x - 28$
   $= x^2 + 3x - 28$

c) $(x - 3)(x - 5) = x \times x + x \times (-5) + (-3) \times x + (-3) \times (-5)$
   $= x^2 - 5x - 3x + 15$
   $= x^2 - 8x + 15$

d) $(2x + 3)(5x - 9) = (2x) \times (5x) + (2x) \times (-9) + 3 \times (5x) + 3 \times (-9)$
   $= 10x^2 - 18x + 15x - 27$
   $= 10x^2 - 3x - 27$

EXAM TIP

Remember: the **FOIL** method means 'First – Outside – Inside – Last'.

2.2 Manipulating algebraic expressions

Extended

WORKED EXAMPLE

Expand and simplify

$(x - 1)(x + 2)(x + 4)$ [3 marks]

$(x - 1)(x + 2)(x + 4)$
$= (x^2 - x + 2x - 2)(x + 4)$ — Expand and simplify the first pair of brackets.
$= (x^2 + x - 2)(x + 4)$
$= x^3 + 4x^2 + x^2 + 4x - 2x - 8$ — Multiply the quadratic expression by the third bracket.
$= x^3 + 5x^2 + 2x - 8$

KEY SKILLS

You need to be able to expand three brackets.

WORKED EXAMPLE

Factorise these expressions.

a) $9ab + 3b^2$ [1 mark]
b) $9x + 5x + 18y + 10y$ [1 mark]
c) $9r + 4s + 5r + 3s$ [1 mark]

a) $9ab$ and $3b^2$ have an HCF of $3b$.
$9ab + 3b^2 = 3b(3a + b)$.
b) $9x + 5x + 18y + 10y = 14x + 28y$ and $14x$ and $28y$ have an HCF of 14.
$9x + 5x + 18y + 10y = 14x + 28y = 14(x + 2y)$
c) $9r + 4s + 5r + 3s = 14r + 7s$ and $14r$ and $7s$ have an HCF of 7.
$9r + 4s + 5r + 3s = 14r + 7s = 7(2r + s)$

KEY SKILLS

You must be able to factorise algebraic expressions.

EXAM TIP

Expanding expressions and factorising expressions are the opposites of each other.

Extended

WORKED EXAMPLE

Factorise:

a) $x^2 + 7x + 6$
b) $x^2 + 3x - 28$
c) $x^2 - 7x + 12$
d) $x^2 - 2x - 15$ [4 marks]

a) $x^2 + 7x + 6$ Find two numbers that multiply to give 6 and add up to 7.
These are 6 and 1.
Therefore $x^2 + 7x + 6 = (x + 6)(x + 1)$.

KEY SKILLS

You must be able to factorise quadratic expressions of the forms $a^2 + 2ab + b^2$ and $ax^2 + bx + c$.

EXAM TIP

When you have factorised an expression, expand the expression again, if you have time, to check that you got it right.

55

b) $x^2 + 3x - 28$ Find two numbers that multiply to give −28 and add up to 3.

These are 7 and −4.

Therefore $x^2 + 3x - 28 = (x + 7)(x - 4)$.

c) $x^2 - 7x + 12$ Find two numbers that multiply to give 12 and add up to −7.

These are −3 and −4.

Therefore $x^2 - 7x + 12 = (x - 3)(x - 4)$.

d) $x^2 - 2x - 15$ Find two numbers that multiply to give −15 and add up to −2.

These are 3 and −5.

Therefore $x^2 - 2x - 15 = (x + 3)(x - 5)$.

WORKED EXAMPLE

Factorise:

a) $x^2 - 16$ [1 mark]

b) $x^2 + 7x$ [1 mark]

a) This is the difference of two squares, so $a^2 - b^2 = (a + b)(a - b)$.

In this case, $a = x$ and $b = 4$; therefore $x^2 - 16 = (x + 4)(x - 4)$.

b) $x^2 + 7x = x(x + 7)$

👍 EXAM TIP

If there is no middle term, it is called 'the difference of two squares'.

WORKED EXAMPLE

Factorise $2x^2 + 7x + 6$. [2 marks]

Step 1 The brackets must be of the form $(2x + \ )(x + \ )$ because 2 is prime.

Step 2 The missing numbers must multiply to give 6, so either 1 and 6 or 2 and 3.

Step 3 Try the four possible combinations.

$(2x + 1)(x + 6)$ gives x term $12x + x = 13x$

$(2x + 6)(x + 1)$ gives x term $2x + 6x = 8x$

$(2x + 2)(x + 3)$ gives x term $6x + 2x = 8x$

$(2x + 3)(x + 2)$ gives x term $4x + 3x = 7x$

Only $(2x + 3)(x + 2)$ gives the $7x$ term in the middle, so $2x^2 + 7x + 6 = (2x + 3)(x + 2)$

👍 EXAM TIP

Factorising an equation in this way is sometimes called factorising 'by inspection'.

2.2 Manipulating algebraic expressions

? QUESTIONS

1. Expand these expressions.
   a) $7(p + 3q)$
   b) $6(5m - 7n)$
   c) $3(5a + 2b) - 6(2a - 3b)$

2. Multiply out the brackets in these expressions.
   a) $(2x + 1)(3x + 2)$
   b) $(5x + 2)(3x + 4)$
   c) $(6t - 1)(2t - 3)$
   d) $(2y - 9)(3y - 1)$
   e) $(7z - 1)(2z + 3)$
   f) $(9r - 2)(3r + 2)$
   g) $(7e - 11)(2e + 1)$
   h) $(8q + 1)(5q - 3)$
   i) $(3p - 1)(3p + 1)$
   j) $(7y + 2)(7y - 2)$
   k) $(2k + 1)(k + 3)$
   l) $(2v - 1)(5v + 1)$

3. Multiply out the brackets and simplify these expressions.
   a) $(x + 2)(x + 3)$
   b) $(x + 5)(x + 4)$
   c) $(t + 1)(t + 2)$
   d) $(3q + 1)(2q - 1)$
   e) $(5y + 2)(2y - 3)$
   f) $(5m - 1)(5m + 1)$
   g) $(2y + 1)(2y - 1)$
   h) $(3p + 2)^2$
   i) $(2q - 1)^2$
   j) $(5d + 2e)(2d - 3e)$
   k) $(5p + 3q)(4p + q)$
   l) $(7s - 3t)(2s - t)$

4. Expand and simplify these expressions.
   a) $(x + 3)^2$
   b) $(y + 5)^2$
   c) $(y - 4)^2$
   d) $(z - 6)^2$
   e) $(2w - 3)^2$
   f) $(5t - 2)^2$
   g) $(3a + b)(2a + b)$
   h) $(3m - 2n)(5m - n)$
   i) $(5p + 2q)(3p - 4q)$
   j) $(2x - 3y)(5x - 2y)$
   k) $(3c + 2d)^2$
   l) $(5p - 3q)^2$

5. Factorise these expressions.
   a) $2x - 4xy$
   b) $3xy + 4yz$
   c) $12x^2 + 14xy$
   d) $16abc - 24a2b$
   e) $8de^3 - 8d^3e$
   f) $35a^2bc - 21ab + 7bc^2$

Extended

6. Factorise these quadratic expressions.
   a) $x^2 + 9x + 18$
   b) $x^2 - x - 20$
   c) $x^2 - 7x + 10$
   d) $x^2 + 3x - 40$
   e) $x^2 - x - 42$
   f) $x^2 + 7x + 12$
   g) $x^2 + 2x - 24$
   h) $x^2 - 16$
   i) $x^2 + 3x$
   j) $x^2 - 25$

7. Factorise these quadratic expressions.
   a) $x^2 - 5x - 6$
   b) $x^2 + 5x + 6$
   c) $x^2 + 5x - 6$
   d) $x^2 - 5x + 6$
   e) $x^2 - 4x - 60$
   f) $x^2 + 5x - 36$
   g) $x^2 - 20x + 99$
   h) $x^2 - 1$
   i) $x^2 + x - 132$
   j) $x^2 + 6x + 9$
   k) $x^2 - 10x + 25$
   l) $x^2 - 100$

8. Factorise these quadratic expressions.
   a) $x^2 + 7x + 12$
   b) $x^2 + 7x + 10$
   c) $x^2 - 5x - 6$
   d) $x^2 - 5x + 6$
   e) $2x^2 + 5x - 12$
   f) $3x^2 + 11x + 6$
   g) $4x^2 + 12x + 5$
   h) $5x^2 + 13x + 8$

9. Factorise these quadratic expressions.
   a) $a^2 + 5ab$
   b) $r^2 + 2r$
   c) $t^2 - 36$
   d) $b^2 + 11b + 24$
   e) $4p^2 + 20p + 9$
   f) $5q^2 - 8q - 4$

10. Factorise these quadratic expressions.
    a) $x^2 + 3x + 2$
    b) $y^2 - 9$
    c) $z^2 + 2z$
    d) $n^2 - n - 6$
    e) $4p^2 - 8p - 5$
    f) $3q^2 - 8q + 4$

11. Factorise these quadratic expressions.
    a) $x^2 - 9x + 20$
    b) $x^2 - 3x - 10$
    c) $4x^2 - 11x + 6$
    d) $6x^2 - 13x - 5$
    e) $8x^2 - 13x + 5$
    f) $6x^2 + 17x + 5$

12. Expand and simplify:
    a) $(x + 1)(x - 2)(x + 3)$
    b) $(x - 3)(x + 5)(x - 6)$
    c) $(2x + 1)(x - 5)(x + 4)$
    d) $(3x - 1)(2x + 3)(4x - 5)$

To **Raise your grade** now try questions 10 and 16 on pages 95–96

2 Algebra and graphs

2.3 Algebraic fractions

YOU NEED TO:
- **E** Manipulate algebraic fractions.
- **E** Factorise and simplify rational expressions.

Extended

> **◀◀ RECAP**
> An algebraic fraction is a fraction that contains algebraic expressions.

WORKED EXAMPLE

Simplify $\dfrac{x+1}{3} + \dfrac{x-3}{4}$. [2 marks]

LCM of 3 and 4 is 12 so express both terms as fractions with denominator 12.

So

$\dfrac{x+1}{3} = \dfrac{4x+4}{12}$ and $\dfrac{x-3}{4} = \dfrac{3x-9}{12}$

Therefore

$\dfrac{x+1}{3} + \dfrac{x-3}{4} = \dfrac{4x+4}{12} + \dfrac{3x-9}{12} = \dfrac{7x-5}{12}$

> 🔑 **KEY SKILLS**
> You must be able to manipulate algebraic fractions.

> 👍 **EXAM TIP**
> The main thing to remember is that, with algebra, you are allowed to do only the things that you are allowed to do with actual numbers.

WORKED EXAMPLE

Simplify $\dfrac{3(4x-1)}{2} - \dfrac{2(5x+3)}{3}$. [2 marks]

LCM of 2 and 3 is 6 so express both terms as fractions with denominator 6.

So

$\dfrac{3(4x-1)}{2} = \dfrac{36x-9}{6}$ and $\dfrac{2(5x+3)}{3} = \dfrac{20x+12}{6}$

Therefore

$\dfrac{3(4x-1)}{2} - \dfrac{2(5x+3)}{3} = \dfrac{36x-9}{6} - \dfrac{20x+12}{6} = \dfrac{16x-21}{6}$

58

2.3 Algebraic fractions

WORKED EXAMPLE

Simplify $\dfrac{3}{1-x} - \dfrac{2}{1+x}$. [2 marks]

LCM of $(1-x)$ and $(1+x)$ is $(1-x)(1+x)$ so

$$\dfrac{3}{1-x} = \dfrac{3(1+x)}{(1-x)(1+x)} = \dfrac{3+3x}{(1-x)(1+x)}$$

and

$$\dfrac{2}{1+x} = \dfrac{2(1-x)}{(1+x)(1-x)} = \dfrac{2-2x}{(1+x)(1-x)}$$

So

$$\dfrac{3}{1-x} - \dfrac{2}{1+x}$$

$$= \dfrac{3+3x}{(1-x)(1+x)} - \dfrac{2-2x}{(1+x)(1-x)}$$

$$= \dfrac{1+5x}{(1+x)(1-x)}$$

WORKED EXAMPLE

Factorise and simplify:

a) $\dfrac{x^2 - 4x}{x^2 - x - 12}$ [2 marks]

b) $\dfrac{x^2 - 7x + 12}{2x^2 - 7x + 3}$ [2 marks]

a) $\dfrac{x^2 - 4x}{x^2 - x - 12}$

$x^2 - 4x = x(x-4)$

$x^2 - x - 12 = (x-4)(x+3)$

Hence

$$\dfrac{x^2 - 4x}{x^2 - x - 12} = \dfrac{x\cancel{(x-4)}}{\cancel{(x-4)}(x+3)}$$

$$= \dfrac{x}{x+3}$$

b) $\dfrac{x^2 - 7x + 12}{2x^2 - 7x + 3}$

$x^2 - 7x + 12 = (x-3)(x-4)$

$2x^2 - 7x + 3 = (x-3)(2x-1)$

Hence

$$\dfrac{x^2 - 7x + 12}{2x^2 - 7x + 3} = \dfrac{\cancel{(x-3)}(x-4)}{\cancel{(x-3)}(2x-1)}$$

$$= \dfrac{x-4}{2x-1}$$

🔑 KEY SKILLS

You must be able to factorise and simplify rational expressions.

👍 EXAM TIP

When you see a question like this, you should expect the expressions on the top and bottom to have a bracket in common when factorised. This is often (although not always) the case, but when it is, it will help you to factorise the second expression more quickly.

2 Algebra and graphs

❓ QUESTIONS

1. Simplify these expressions, leaving your answers as fractions in their simplest form.

   a) $\dfrac{2x+1}{3} + \dfrac{x+1}{5}$ b) $\dfrac{2x-5}{2} + \dfrac{x+3}{3}$

   c) $\dfrac{2x+3}{4} + \dfrac{3x-2}{6}$ d) $\dfrac{3x+1}{2} + \dfrac{x+2}{7}$

   e) $\dfrac{3x+5}{3} - \dfrac{x+2}{4}$ f) $\dfrac{5x+1}{5} - \dfrac{3x-2}{6}$

   g) $\dfrac{x-1}{6} - \dfrac{5x+3}{12}$ h) $\dfrac{6x-1}{4} - \dfrac{2x-5}{5}$

2. Simplify these expressions, leaving your answers as fractions in their simplest form.

   a) $\dfrac{x}{2} + \dfrac{3x+1}{5} + \dfrac{2x-1}{10}$

   b) $\dfrac{x-1}{4} + \dfrac{4x+1}{5} + \dfrac{2x+3}{20}$

   c) $\dfrac{x-3}{2} + \dfrac{2x-1}{3} + \dfrac{x+1}{6}$

   d) $x + \dfrac{x}{2}$

   e) $x + \dfrac{2x+1}{2} + \dfrac{x}{3}$

   f) $x + 1 + \dfrac{3x+2}{3} + \dfrac{x-2}{4}$

3. Simplify these expressions, leaving your answers as fractions in their simplest form.

   a) $\dfrac{2}{x+1} + \dfrac{3}{x+2}$ b) $\dfrac{3}{x+2} + \dfrac{4}{x+3}$

   c) $\dfrac{5}{x+1} + \dfrac{7}{x-1}$ d) $\dfrac{5}{x+3} - \dfrac{2}{x+2}$

   e) $\dfrac{3}{2x+3} + \dfrac{4}{3x-1}$ f) $\dfrac{5}{3x+1} + \dfrac{4}{4x-3}$

4. Simplify these fractions.

   a) $\dfrac{x^2 + 3x + 2}{x + 2}$ b) $\dfrac{x^2 + 5x + 6}{x + 3}$

   c) $\dfrac{2x^2 + 3x + 1}{x + 1}$ d) $\dfrac{x^2 + x - 12}{x + 4}$

   e) $\dfrac{x^2 - 7x + 10}{x - 2}$ f) $\dfrac{4x^2 - 8x + 3}{2x - 3}$

   g) $\dfrac{x^2 - 1}{x - 1}$ h) $\dfrac{25x^2 - 1}{5x + 1}$

To **Raise your grade** now try question 14 on page 96

2.4 Indices

YOU NEED TO:
- Understand and use indices (positive, zero and negative).
- Understand and use the rules of indices.
- **E** Understand and use fractional indices.

◀◀ RECAP

When you write $2 \times 2 \times 2 \times 2 \times 2$ as 2^5 you are using index notation.

When the index is a positive whole number, such as 5, 2^5 means 'five 2s multiplied together'.

2 is the base → 2^5 ← 5 is the index

◀◀ RECAP

The three laws of indices are:
- $a^m \times a^n = a^{m+n}$
- $a^m \div a^n = a^{m-n}$
- $(a^m)^n = a^{mn}$

WORKED EXAMPLE

Simplify these expressions.

a) $a^2 \times a^3$
b) $(p^3)^4$
c) $\dfrac{q^{11}}{q^2}$
d) $\dfrac{r^4}{r^7}$ [4 marks]

a) $a^2 \times a^3 = a^5$
b) $(p^3)^4 = p^{12}$
c) $q^{11} \div q^2 = q^9$
d) $r^4 \div r^7 = r^{4-7} = r^{-3}$

KEY SKILLS

You must be able to use the rules of indices.

WORKED EXAMPLE

Simplify $\dfrac{(3x^2y)^2 \times (2x^3y^2)^3}{(2x^5y^7)^2}$ [3 marks]

$\dfrac{(3x^2y)^2 \times (2x^3y^2)^3}{(2x^5y^7)^2} = \dfrac{9x^4y^2 \times 8x^9y^6}{4x^{10}y^{14}}$

$= \dfrac{72x^{13}y^8}{4x^{10}y^{14}}$

$= 18x^3y^{-6}$

$= \dfrac{18x^3}{y^6}$

EXAM TIP

Remember to apply the power of each bracket to the numbers as well as the letters.

Set out your working step by step and use only one rule of indices on each line.

You can use the rule for negative indices to write your final answer as a fraction.

2 Algebra and graphs

Extended

WORKED EXAMPLE

Simplify these expressions.

a) $\left(\sqrt{a}\right)^{-4}$ [1 mark]

b) $2a^{\frac{1}{2}}b^{\frac{2}{3}} \times 3a^{-\frac{3}{2}}b^{\frac{2}{3}}$ [3 marks]

a) $\left(a^{\frac{1}{2}}\right)^{-4} = a^{-2}$

b) $2a^{\frac{1}{2}}b^{\frac{2}{3}} \times 3a^{-\frac{3}{2}}b^{\frac{2}{3}} = 2 \times 3 \times a^{\frac{1}{2}} \times a^{-\frac{3}{2}} \times b^{\frac{2}{3}} \times b^{\frac{2}{3}}$

$= 6a^{-1}b^{\frac{4}{3}}$

KEY SKILLS

You must be able to use and interpret fractional indices.

EXAM TIP

Reorder the terms and then use the rules of indices.

QUESTIONS

1. Simplify these expressions, leaving your answers as fractions in their simplest form.
   a) $a^2 \times a^3$
   b) $b^3 \times b^{14}$
   c) $c^6 \times c^5$
   d) $x^6 \times x^7$
   e) $y^9 \div y^2$
   f) $z^{11} \div z^4$
   g) $(q^3)^4$
   h) $(w^2)^6$
   i) $(r^2)^3 \div r^4$
   j) $(s^5)^3 \times s^6$
   k) $\dfrac{t^2 \times t^5}{t^3}$
   l) $\dfrac{(w^3)^4 \times w^6}{w^2}$

2. Find a when:
   a) $2^3 \times 2^a = 2^7$
   b) $x^5 \times x^a = x^6$
   c) $y^2 \times y^a = y^2$
   d) $r^a \div r^{11} = r^9$
   e) $s^a \div s^9 = s^7$
   f) $2^9 \div 2^a = 2^2$
   g) $(3^5)^a = 3^{20}$
   h) $(u^a)^a = u^9$
   i) $(x^a)^{a+1} = x^{42}$

3. Simplify these fractions.
   a) $\dfrac{(2x^8) \times (6x^5)}{4x^9}$
   b) $\dfrac{(4y^6) \times (5y^7)}{10y^{11}}$
   c) $\dfrac{(8z^2) \times (3z^3)}{2z^4}$
   d) $\dfrac{(6a^3) \times (4a^9)}{(2a^2) \times (3a^8)}$
   e) $\dfrac{(12c^3) \times (4c^9)}{(2c^2) \times (3c^6)}$
   f) $\dfrac{(10h^2) \times (18h^5)}{(5h^4) \times (3h)}$

4. Simplify these fractions.
   a) $\dfrac{(4a^3b^4) \times (10a^2b^5)}{5a^4b^2}$
   b) $\dfrac{(6mn^2) \times (10m^3n^4)}{4mn^2}$
   c) $\dfrac{(2r^2t^4) \times (10r^5t^3)}{5r^2t^5}$
   d) $\dfrac{(3g^2h^5) \times (2g^3h^6)}{6g^3h}$

5. Simplify these expressions.
   a) $(2x^2y)^3$
   b) $(3x^3y^4)^2$
   c) $(8x^2y^3)^2$
   d) $(5x^2y^5)^3$
   e) $(9x^6y^7)^2$
   f) $(8x^7y^3)^2$
   g) $(2x^2y^3)^4 \times (3x^2y^3)^3$
   h) $(2x^3y^5)^3 \times (5x^3y^4)^2$

Extended

6. Simplify these expressions.
   a) $(\sqrt[3]{x})^{-4}$
   b) $\sqrt{y^{-7}}$
   c) $\sqrt[5]{p^6 \times p^{-7}}$

7. Simplify these expressions.
   a) $4x^{\frac{1}{2}} \times 3x^{-\frac{1}{3}}$
   b) $6x^{\frac{1}{3}}y^{\frac{1}{2}} \times 5x^{-1}y^{\frac{2}{5}}$
   c) $\dfrac{8x^{\frac{4}{5}}y^{\frac{1}{3}}}{4x^{-\frac{1}{5}}y^{-\frac{1}{3}}}$

To **Raise your grade** now try questions 1 and 8 on page 95

2.5 Equations and formulas

> **YOU NEED TO:**
> - Construct expressions, equations and formulas.
> - Solve linear equations in one unknown.
> - **E** Solve fractional equations with numerical and linear algebraic denominators.
> - Solve simultaneous linear equations in two unknowns.
> - **E** Solve simultaneous equations, involving one linear and one non-linear.
> - **E** Solve quadratic equations by factorisation, completing the square and by use of the quadratic formula.
> - Change the subject of formulas.

◀◀ RECAP

In a linear equation, the highest power of x is 1.

For example, $3x + 2 = 17$ is a linear equation.

When you solve the equation you find the value of x that makes the left-hand side of the equation equal to the right-hand side.

The solution of $3x + 2 = 17$ is $x = 5$ since $3 \times 5 + 2 = 17$.

You solve an equation by doing the same operations to both sides.

🔑 KEY SKILLS

You must be able to derive and solve simple linear equations in one unknown.

🔑 KEY SKILLS

You must be able to derive and solve simultaneous linear equations in two unknowns.

WORKED EXAMPLE

Solve the equation $3x - 5 = 16$. **[2 marks]**

$3x - 5 = 16$	Add 5 to both sides.
$3x = 21$	Divide both sides by 3.
$x = 7$	

👍 EXAM TIP

As a general rule, to solve a linear equation:

Step 1 Clear the fractions.
Step 2 Expand the brackets.
Step 3 Put all the x-terms on one side of the equation.
Step 4 Simplify the equation.
Step 5 Check your solution.

WORKED EXAMPLE

Solve the equation $2(x + 4) = 14$. **[2 marks]**

$2(x + 4) = 14$	Expand the brackets.
$2x + 8 = 14$	Subtract 8 from both sides.
$2x = 6$	Divide both sides by 2.
$x = 3$	

63

2 Algebra and graphs

WORKED EXAMPLE

Solve the equation $\frac{3x}{4} = 12$. **[2 marks]**

$\frac{3x}{4} = 12$ — Clear the fractions by multiplying both sides by 4.

$3x = 48$ — Divide both sides by 3.

$x = 16$

WORKED EXAMPLE

Apples cost a cents and bananas cost b cents.

Anvi buys 3 apples and 1 banana and pays 90 cents.

Riya buys 1 apple and 2 bananas and pays 80 cents.

a) Derive two simultaneous equations from this information. **[2 marks]**
b) Solve the simultaneous equations to find a and b. **[3 marks]**

---

a) The cost of 3 apples = $3a$ cents and the cost of 1 banana = b cents.

So, from Anvi, $3a + b = 90$ (equation 1)

The cost of 1 apple = a cents and the cost of 2 bananas = $2b$ cents.

So, from Riya, $a + 2b = 80$ (equation 2)

b) You can solve these simultaneous equations by **elimination**.

Multiply (equation 1) by 2 so that both equations have a $2b$ term.

Give this new equation the label (equation 3).

$6a + 2b = 180$ (equation 3)

Subtract equation (2) from equation (3) to eliminate the $2b$ terms.

$(6a + 2b) - (a + 2b) = 180 - 80$

$5a = 100$, so $a = 20$

Use the value of a in (equation 1) to find b.

$3 \times 20 + b = 90$, so $b = 30$

Therefore, an apple costs 20 cents and a banana costs 30 cents.

👍 **EXAM TIP**

They are called **simultaneous** equations because the solutions satisfy both of the equations at the same time. It is a good idea to label your simultaneous equations with a number, so you can keep track of which one is which as you change and combine them.

The elimination method works by multiplying one or both of the equations by a number (different numbers if you are multiplying both equations), so that the size of the coefficient of one of the variables is the same in both equations.

When using the elimination method, you aim to get the equations into one of these forms:

Same coefficients and same signs   or   **Same coefficients but different signs**

$$3x + 2y = 19 \quad (1)$$
$$6x + 2y = 34 \quad (2)$$

$$2x + 6y = 26 \quad (1)$$
$$7x - 6y = 10 \quad (2)$$

SAME SIGNS SUBTRACT

In this case, **subtract** the equations: (2) − (1)

$$6x - 3x + 2y - 2y = 34 - 19$$
$$3x = 15$$
$$x = 5$$

Put $x = 5$ in equation (1).

$$3 \times 5 + 2y = 19$$
$$15 + 2y = 19$$
$$2y = 4$$
$$y = 2$$

DIFFERENT SIGNS ADD

In this case, **add** the equations: (1) + (2)

$$2x + 7x + 6y - 6y = 26 + 10$$
$$9x = 36$$
$$x = 4$$

Put $x = 4$ in equation (1).

$$2 \times 4 + 6y = 26$$
$$8 + 6y = 26$$
$$6y = 18$$
$$y = 3$$

WORKED EXAMPLE

Solve the simultaneous equations $2x + 5y = 19$ and $y = x + 1$ **[3 marks]**

First, number your two equations.

$$2x + 5y = 19 \quad (1)$$
$$y = x + 1 \quad (2)$$

Equation (2) gives y in terms of x, so substitute $(x + 1)$ for y in equation (1).

$$2x + 5(x + 1) = 19$$
$$7x + 5 = 19$$
$$7x = 14$$
$$x = 2$$

Put $x = 2$ in equation (2):

$$y = 2 + 1 = 3$$

The solution is $x = 2, y = 3$.

👍 EXAM TIP

The elimination method is the quickest method, when it works. Often though, you will need to use substitution.

Extended

WORKED EXAMPLE

 Solve the simultaneous equations $y = x^2 + 3$ and $y = 10 - \dfrac{3}{2}x$ **[3 marks]**

Because both equations start with $y =$, you can equate the right-hand sides to give

$$x^2 + 3 = 10 - \dfrac{3}{2}x$$

Multiply both sides by 2: $2x^2 + 6 = 20 - 3x$

🔑 KEY SKILLS

You must be able to derive and solve simultaneous equations, involving one linear and one quadratic equation.

2 Algebra and graphs

→ Bring everything to one side: $2x^2 + 3x - 14 = 0$

Factorise: $(2x + 7)(x - 2) = 0$

Either $2x + 7 = 0$, so $x = -\dfrac{7}{2}$ or

$x - 2 = 0$, so $x = 2$

But $y = x^2 + 3$, so when $x = 2$, $y = 2^2 + 3 = 7$

and when $x = -\dfrac{7}{2}$, $y = \left(-\dfrac{7}{2}\right)^2 + 3 = \dfrac{49}{4} + 3 = 15.25$

✎ APPLY

If one equation is quadratic and one is linear, what are the maximum and minimum number of solutions the simultaneous equations could have?

◀◀ RECAP

Quadratic equations are equations where the highest power of any variable is 2. There are three main ways to solve them.

🔑 KEY SKILLS

You must be able to derive and solve quadratic equations by factorisation, completing the square and by use of the formula.

WORKED EXAMPLE

Solve the equation $x^2 + 5x - 24 = 0$. **[3 marks]**

This can be solved by factorising.

$x^2 + 5x - 24 = 0$

$(x + 8)(x - 3) = 0$

Either $x + 8 = 0$ giving $x = -8$

or $x - 3 = 0$ giving $x = 3$

👍 EXAM TIP

Factorising is always the quickest and easiest method when the equation is easy to factorise. Always start by seeing if this is possible.

WORKED EXAMPLE

Solve $x^2 + 8x + 5 = 0$ by completing the square. **[3 marks]**

$x^2 + 8x + 5 = (x + 4)^2 - 11$

$8 \div 2 = 4 \qquad -4^2 + 5 = -11$

So you need to solve $(x + 4)^2 - 11 = 0$

Hence $(x + 4)^2 = 11$

So $x + 4 = \pm\sqrt{11}$ and $x = -4 \pm \sqrt{11}$.

$x = -0.683$ or $x = -7.32$ (to 3 s.f.)

👁 WATCH OUT!

Don't forget the 'plus or minus' when you square root both sides.

👍 EXAM TIP

Completing the square is most useful when the coefficient of x^2 is 1 and the coefficient of x is an even number. If this is not the case, it can become messy, and you might want to just use the quadratic formula.

◀◀ RECAP

The quadratic formula states that the solutions to the equation

$ax^2 + bx + c = 0$ are $x = \dfrac{-b + \sqrt{b^2 - 4ac}}{2a}$ and $x = \dfrac{-b - \sqrt{b^2 - 4ac}}{2a}$

The two solutions are usually combined in the form $x = \dfrac{-b \pm \sqrt{b^2 - 4ac}}{2a}$ where ± means 'plus or minus'.

To use this method, replace the letters a, b and c with the numbers that come from the particular equation you want to solve.

2.5 Equations and formulas

WORKED EXAMPLE

 Solve the equation $3x^2 - 2x - 7 = 0$. **[3 marks]**

The equation can't be factorised, and the coefficient of x^2 is not 1, so solve it using the quadratic formula.

$$x = \frac{-b \pm \sqrt{b^2 - 4ac}}{2a}$$

Write down the values of a, b and c.

$a = 3$, $b = -2$, $c = -7$

Substitute these values in the formula.

$$x = \frac{-2 \pm \sqrt{(-2)^2 - 4 \times 3 \times -7}}{6}$$

$$= \frac{2 \pm \sqrt{88}}{6}$$

$$= \frac{2 \pm 9.3808...}{6}$$

$x = 1.90$ or $x = -1.23$ to 3 s.f.

> **WATCH OUT!**
>
> It is probably not a good idea to type the whole formula into the calculator in one go.
>
> Instead, break the calculation into smaller parts.

WORKED EXAMPLE

Make x the subject of the formula $2x + 3b = c$. **[2 marks]**

Step 1 Clear the fractions. *No fractions*

Step 2 Multiply out any brackets involving x. *No brackets to multiply out.*

Step 3 Put x-terms together on one side.

$2x = c - 3b$

Step 4 Divide by the coefficient of the x-term. *Coefficient of x-term is 2.*

$$x = \frac{c - 3b}{2}$$

> **KEY SKILLS**
>
> You must be able to rearrange simple formulas.

WORKED EXAMPLE

Make x the subject of the formula $\dfrac{ax + b}{x} = c$.

Step 1 Clear the fractions.

$ax + b = c \times x = cx$

Step 2 Multiply out any brackets involving x. *No brackets to multiply out.*

Step 3 Put x-terms on one side and write as $x() =$

$cx - ax = b$

$x(c - a) = b$

Step 4 Divide by the coefficient of the x-term.

$$x = \frac{b}{c - a}$$

2 Algebra and graphs

WORKED EXAMPLE

Make x the subject of the formula $\dfrac{a}{x} + b = c$.

Step 1 Clear the fractions.

$a + bx = cx$

Step 2 Multiply out any brackets involving x. No brackets.

Step 3 Put x-terms on one side and write as $x() =$

$a = cx - bx$

$a = x(c - b)$

Step 4 Divide by the coefficient of the x-term.

$x = \dfrac{a}{c - b}$

EXAM TIP

Here x appears twice, so factorise to give $x(c - b) = a$.

WORKED EXAMPLE

Make x the subject of the formula $a = \dfrac{bx^2 + c}{d}$.

Step 1 Clear the fractions.

$ad = bx^2 + c$

Step 2 Multiply out any brackets involving x^2. No brackets.

Step 3 Put x^2 term on one side.

$ad - c = bx^2$

Step 4 Divide by the coefficient of the x^2 term.

$x^2 = \dfrac{ad - c}{b}$

Step 5 Take square roots.

$x = \pm\sqrt{\dfrac{ad - c}{b}}$

EXAM TIP

If the equation involves x^2 then, first of all, make x^2 the subject and then take the square root of both sides to find x. Include the ± sign when taking square roots.

QUESTIONS

1. The three angles in a triangle are a, $a + 20$ and $a + 25$.

   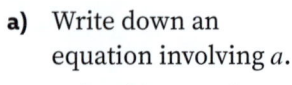

   a) Write down an equation involving a.
   b) Solve this equation to find a.

2. Three consecutive whole numbers add up to 144. If the lowest number is n:
   a) write down expressions for the other two numbers in terms of n
   b) write down an equation involving n
   c) find n and hence find the other two numbers.

3. Solve the following equations.
   a) $3(x - 1) = 2(x + 1)$
   b) $2(5x + 2) = 6(3x - 2)$
   c) $5(7x - 3) = 4(9x - 4)$

4. Solve the following equations.
   a) $\dfrac{15}{x} = 5$
   b) $\dfrac{6}{x} = 2$
   c) $\dfrac{42}{x + 3} = 6$
   d) $\dfrac{35}{2x + 1} = 5$

5. Solve these simultaneous equations by elimination.

   a) $2x + y = 11$
      $3x - y = 14$

   b) $3u + 2v = 10$
      $7u - v = 29$

   c) $11p + 3q = 71$
      $5p - q = 37$

   d) $9a + 2b = 41$
      $5a - 4b = 33$

   e) $7p - 3q = 15$
      $5p + 2q = 19$

   f) $13b - 7c = 47$
      $7b - 9c = 41$

6. Solve these simultaneous equations (by the substitution method).

   a) $5m + 3n = 27$
      $m = 7 - n$

   b) $2p + 7q = 3$
      $p = 6 + q$

   In parts **c)** to **f)**, first rearrange one of the equations.

   c) $u + 2v = 7$
      $2u + 3v = 11$

   d) $3p + 2q = 21$
      $p - 3q + 4 = 0$

   e) $7r + 2s = 17$
      $r - 3s = 9$

   f) $5x - 7y + 5 = 26$
      $x - y - 9 = 0$

7. The total cost of 5 first class tickets and 2 second class tickets is €246. The total cost of 2 first class and 3 second class tickets is €149.

   Let the price of a first class ticket be €x and the price of a second class ticket be €y.

   a) Write down a pair of simultaneous equations involving x and y.
   b) Find x and y.

8. Solve these equations.

   a) $x^2 + 7x + 12 = 0$
   b) $y^2 + 13y + 22 = 0$
   c) $m^2 - 5m - 6 = 0$
   d) $a^2 - 5a + 6 = 0$
   e) $z^2 - 4z - 12 = 0$
   f) $z^2 + 2z + 1 = 0$
   g) $c^2 - 15c + 36 = 0$
   h) $t^2 - 18t + 81 = 0$
   i) $r^2 - 6r = 0$
   j) $t^2 + 11t = 0$
   k) $w^2 - 3w = 0$
   l) $k^2 + k = 0$

9. Solve the following equations by completing the square (leaving square roots in your answers).

   a) $x^2 + 2x - 1 = 0$
   b) $x^2 - 4x - 3 = 0$
   c) $x^2 + 12x + 36 = 0$
   d) $x^2 + 20x + 5 = 0$
   e) $x^2 + 8x - 9 = 0$
   f) $x^2 - 2x - 7 = 0$

10. Solve the following equations by completing the square. First write them in the form $x^2 + px + q = 0$. Leave square roots in your answers.

    a) $2x^2 + 4x - 6 = 0$
    b) $3x^2 + 15x - 12 = 0$
    c) $2x^2 + 10x + 1 = 0$
    d) $2x^2 + 8x - 12 = 0$

11. a) If $f(x) = x^2 + 4x + 5$. Then show that the equation can be written as $f(x) = (x + 2)^2 + 1$.
    b) Use the result from part **a)** to explain why $f(x)$ cannot take a value lower than 1.
    c) Use the result from part **a)** to also explain why $f(x)$ takes the minimum value of 1 when $x = -2$.

12. Edible Reptiles claims that its bags of snakes contain 5 more snakes than the bags from Snakes Alive and that they charge 1 cent less per snake than Snakes Alive. Anjana buys a bag from Snakes Alive for $5.00. Dhruv buys a bag of the same snakes for $5.70 from Edible Reptiles.

    a) If n is the number of snakes in a bag from Snakes Alive, then find, in terms of n:
       i) how many snakes are in a bag from Edible Reptiles
       ii) the cost per snake (in cents) at both shops.
    b) Set up an equation involving n and solve it to find n.

13. A rectangular box is 23 cm longer than it is wide. Its diagonal measures 65 cm.

    If x is the width of the box then:

    a) use Pythagoras' theorem to find an expression for the length of the box in terms of x
    b) show that $x^2 + 23x - 1848 = 0$
    c) solve this equation to find the exact value of x.

14. Solve the following pairs of simultaneous equations.

    a) $y = x^2 - 6x + 10$
       $y = x + 4$
    b) $y = 8x^2 - 22x + 6$
       $y = 8x - 7$
    c) $x^2 + y^2 = 25$
       $y = 3x - 5$

15. The formula for the volume, V, of a sphere of radius r is $V = \frac{4}{3}\pi r^3$.

    a) Use this formula to calculate the volume of a sphere (to 3 s.f.) of radius 2 m.
    b) Make r the subject of the formula.
    c) Use part **b)** to calculate the radius (to 3 s.f.) of a sphere of volume 200 mm³.

2 Algebra and graphs

16. Make x the subject of each formula.
  a) $mx + n = p$
  b) $a(x + b) = c$
  c) $\dfrac{x + p}{q} = r$
  d) $\dfrac{p(x + q)}{r} = t$
  e) $\dfrac{h}{x} = u$
  f) $\dfrac{k}{x + b} = w$
  g) $\dfrac{d}{ax + b} = c$
  h) $\dfrac{a}{x} + b = c$

17. Make the given variable the subject of each formula.
  a) $\dfrac{aw + b}{c} = d$ (w)
  b) $\dfrac{y + b}{x + t} = c$ (y)
  c) $\dfrac{z}{a} = \dfrac{b}{c}$ (z)
  d) $\dfrac{a}{h} = b$ (h)
  e) $\dfrac{o}{h} = \dfrac{1}{2}$ (o)
  f) $\dfrac{r}{t} = t$ (r)
  g) $\dfrac{t + b}{a + c} = d$ (t)
  h) $\dfrac{a}{g} = \dfrac{p}{q}$ (g)

18. Make x the subject of each of these formulas.
  a) $r + mx = nx$
  b) $ax + b = cx + d$
  c) $x = \dfrac{d + bx}{a}$
  d) $n - x = \dfrac{m + qx}{p}$
  e) $\dfrac{A}{x} = \dfrac{B}{x} + C$
  f) $\dfrac{ax}{x + b} = c$

19. Make the given variable the subject of each formula.
  a) $\dfrac{A + s}{A} = t$ (A)
  b) $\dfrac{aR}{R + 1} = b$ (R)
  c) $\dfrac{ae + b}{ce + d} = 1$ (e)
  d) $\dfrac{ap}{bp + c} = d$ (p)
  e) $\dfrac{bQ}{Q + d} = c$ (Q)
  f) $\dfrac{\sqrt{ax - b}}{c} = d$ (x)
  g) $\sqrt{\dfrac{my + n}{p}} = q$ (y)
  h) $\dfrac{n - aq}{bq + m} = c$ (q)
  i) $\sqrt{\dfrac{a - bt}{t}} = c$ (t)
  j) $\left(\dfrac{ak}{k + b}\right)^2 = p$ (k)

To **Raise your grade** now try questions 2, 3, 4, 5, 7, 11, 12 and 13 on pages 95–96

2.6 Inequalities

YOU NEED TO:

- Represent and interpret inequalities, including on a number line.
- **E** Construct, solve and interpret linear inequalities.
- **E** Represent and interpret linear inequalities in two variables graphically.
- **E** List inequalities that define a given region.

RECAP

You can solve inequalities in the same way as you solve equations.

You can:

- add and subtract any number from both sides of an inequality
- multiply and divide both sides of an inequality by a positive number.

If you multiply or divide by a negative number you have to reverse the inequality sign e.g. $-3 < 1$ but $3 > -1$. Try to avoid doing this as it is easy to make a mistake.

KEY SKILLS

You must be able to derive and solve linear inequalities.

WORKED EXAMPLE

Solve $x + 2 > 7$. **[1 mark]**

$x + 2 > 7$ Subtract 2 from both sides.
$x > 5$

WORKED EXAMPLE

Solve $-7x > 21$. **[2 marks]**

$-7x > 21$ Add $7x$ to both sides.
$0 > 21 + 7x$ Subtract 21 from both sides.
$-21 > 7x$ Divide both sides by 7.
$-3 > x$
$x < -3$

or

$-7x > 21$ Divide both sides by -7 and reverse inequality sign.
$x < -3$

WATCH OUT!

The second method is quicker but you can easily make a mistake.

2 Algebra and graphs

You can express inequalities on a number line.

$x > -3$ can be expressed as

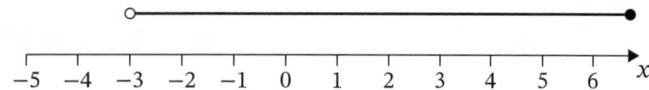

$x \leq 4$ can be expressed as

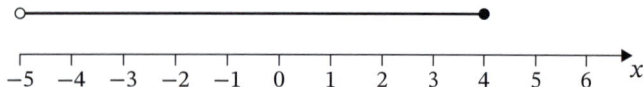

$1 < x \leq 7$ can be expressed as

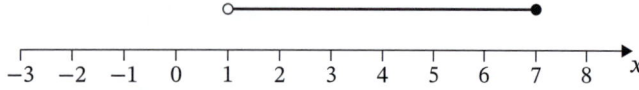

How can you remember whether to fill in the circle or not?

The signs ≥ and ≤ use more ink than > and <.

● uses more ink than ○

'Less than or equal to' and 'greater than or equal to' use more ink than 'less than' or 'greater than'.

≤ goes with 'less than or equal to' and ●

≥ goes with 'greater than or equal to' and ●

< goes with 'less than' and ○

> goes with 'greater than' and ○

Extended

◀◀ RECAP

Just as equations can be represented graphically by lines and curves, inequalities can be represented by regions.

🔑 KEY SKILLS

You must be able to represent inequalities graphically.

WORKED EXAMPLE

By shading the unwanted region on the diagram, show the region defined by the inequalities $x \geq 2$, $y > 1$, $x + y < 6$ **[4 marks]**

First, draw the three lines $x = 2$, $y = 1$, $x + y = 6$

Use dotted lines for $x + y = 6$, and $y = 1$ and a solid line for $x = 2$.

$x \geq 2$ describes the region to the right of $x = 2$. So shade to the left of $x = 2$.

$y > 1$ describes the region above the line $y = 1$. So shade below $y = 1$.

$x + y < 6$ describes the region below the line $x + y = 6$. So shade above $x + y = 6$.

The region $x \geq 2$, $y > 1$, $x + y < 6$ is shown unshaded.

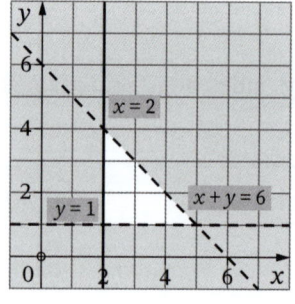

👍 EXAM TIP

Use solid lines to show that the points on the line are included in the region described by an inequality.

Use dotted lines to show that the points on the line are not included in the region described by an inequality.

This is true regardless of whether you are shading the region that is included or that is not included.

2.6 Inequalities

WORKED EXAMPLE

a) Write down the three inequalities that define the unshaded region in the diagram. **[3 marks]**

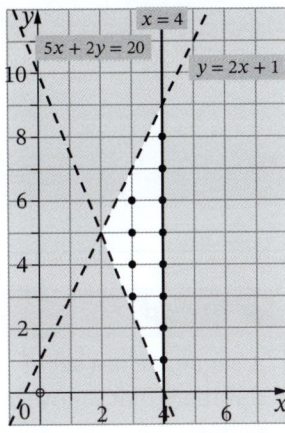

b) Find the maximum value of $x + y$ for points that have integer coordinates in this region. **[2 marks]**

a) The unshaded region is defined by the inequalities $x \leq 4$, $y < 2x + 1$, and $5x + 2y > 20$.

b) In the unshaded region, the points that have integer coordinates are marked as black dots on the diagram.

The table shows the value of $x + y$ at each of the possible points.

x	3	3	3	3	4	4	4	4	4	4	4	4
y	3	4	5	6	1	2	3	4	5	6	7	8
$x + y$	6	7	8	9	5	6	7	8	9	10	11	12

The maximum value of $x + y = 12$.

> **WATCH OUT!**
> Make sure you know whether the question wants you to shade the area that *is* in the region, or the area that *isn't*.

? QUESTIONS

1. Solve the following inequalities.
   a) $3x - 7 < 17$
   b) $10 - 2x \geq 4$
   c) $6x - 3 > 3(x + 2)$
   d) $7 - 4x \leq 9 - x$

   To **Raise your grade** now try questions 2, 4, 5, 7, 12 and 13 on pages 95–96

2.7 Sequences

YOU NEED TO:

- Continue a given number sequence or pattern.
- Recognise patterns in sequences, including the term-to-term rule, and relationships between different sequences.
- Find and use the *n*th term of linear, simple quadratic and simple cubic sequences.
- **E** Find and use the *n*th term of exponential sequences.

WORKED EXAMPLE

Write down the next two terms in each of these sequences:

a) 5, 8, 11, 14, 17, ... [1 mark]
b) 2, 6, 18, 54, ... [1 mark]
c) 22, 15, 8, 1, ... [1 mark]

---

a) The term-to-term rule is to 'add 3', so the next two terms are 20 and 23.
b) The term-to-term rule is to 'multiply by 3', so the next two terms are 162 and 486.
c) The term-to-term rule is to 'subtract 7', so the next two terms are –6 and –13.

WORKED EXAMPLE

Find the formula for the *n*th term of this sequence:

4, 10, 16, 22, 28, ... [2 marks]

The gap between the terms is always 6, therefore it is a linear sequence.

This means that the *n*th term formula is of the form: *n*th term $= an + b$

where a represents the gap between consecutive terms.

nth term $= 6n + b$

When $n = 1$ $4 = 6 \times 1 + b$

therefore $b = -2$

so the formula is nth term $= 6n - 2$

◀◀ RECAP

A sequence is a set of numbers that follows a pattern.

🔑 KEY SKILLS

You must be able to continue a given number sequence.

🔑 KEY SKILLS

You must be able to recognise patterns in sequences, including the term-to-term rule and relationships between different sequences.

🔑 KEY SKILLS

You must be able to find and use the *n*th term of linear sequences.

2.7 Sequences

WORKED EXAMPLE

Find the formula for the nth term of this sequence:

5, 11, 19, 29, 41, ... [3 marks]

The gaps between the terms (the first differences) are 6, 8, 10, 12.

The gaps between these numbers (the second differences) are 2, 2, 2, 2.

As the second differences are equal it means that this is a quadratic sequence.

As the second differences are 2 it means that the formula starts with n^2.

To get from the sequence n^2, which is 1, 4, 9, 16, 25, ...

to the sequence 5, 11, 19, 29, 41, ...

you need to add 4, 7, 10, 13, 16, ... which is the sequence $3n + 1$.

The formula is therefore $n^2 + 3n + 1$.

KEY SKILLS

You must be able to find and use the nth term of quadratic sequences.

EXAM TIP

Whatever the second differences are is double the coefficient of n^2 in the formula.

WORKED EXAMPLE

a) What is the nth term of 19, 13, 7, 1, ...? [2 marks]

b) Which term of 19, 13, 7, 1, ... is equal to −47? [2 marks]

a) The differences are all −6, so the nth term is $-6n + 25$,
which can be written as $25 - 6n$.

b) From part **a)** the nth term is $25 - 6n$, so $25 - 6n = -47$
$$72 = 6n, \text{ so } n = 12$$

So the 12th term is equal to −47.

Extended

WORKED EXAMPLE

Find the nth term formula for this sequence:

3, 5, 9, 17, 33, ... [2 marks]

The sequence is exponential

2^n gives 2, 4, 8, 16, 32, ...

So the nth term formula is $2^n + 1$.

KEY SKILLS

You must be able to find and use the nth term of an exponential sequence.

EXAM TIP

Compare the sequence to a known exponential sequence, in this case, the sequence 2^n.

EXAM TIP

Sometimes the nth term may be referred to as T_n.

75

2 Algebra and graphs

QUESTIONS

1. A football club has 35 000 supporters at its first home match. The attendance increases by 250 at each home game.
   a) How many supporters will be at the nth home game?
   b) If there are 37 750 at the last home game of the season then how many home games did the club play?

2. Joshua decides to save money in the following way.

   He saves $1 in the first week, $1.20 in the second week, $1.40 in the third week, and so on.
   a) How much does he save in the nth week?
   b) How much does he save in the 8th week?
   c) In which week does he first save at least $5?
   d) After 10 weeks, Joshua wants to buy a tennis racquet which costs $19.99.

   He realises that he hasn't saved quite enough. How much more money does he need?

3. Find the first three terms in a sequence whose nth term is:
   a) $5n - 1$
   b) $n^2 - 1$
   c) $\frac{1}{2}n(n+1)$
   d) $\frac{n^3 + 1}{3n}$

4. Find the next two terms in each sequence.
   a) 5, 10, 20, 40, 80, ...
   b) 1, 8, 27, 64, 125, ...
   c) 2, 8, 18, 32, 50, ...
   d) $\frac{3}{4}, \frac{5}{9}, \frac{7}{16}, \frac{9}{25}, \frac{11}{36}, ...$

5. Find the nth term formula for each of these sequences.
   a) 7, 9, 11, 13, ...
   b) 16, 13, 10, 7, ...
   c) 2, 5, 10, 17, 26, ...
   d) 2, 8, 18, 32, 50, ...
   e) 0, 7, 26, 63, 124, ...

6. Find the nth term formula for each of these sequences.
   a) 2, 7, 14, 23, 34, ...
   b) −2, 2, 8, 16, 26, ...
   c) 3, 11, 21, 33, 47, ...
   d) 6, 13, 22, 33, 46, ...

Extended

7. a) Write down the next two terms in each of these sequences.
   i) 4, 6, 10, 18, 34, ...
   ii) 3, 6, 12, 24, 48, ...

   b) Find the nth term formula for each of these sequences.
   i) 1, 3, 7, 15, 31, ...
   ii) 6, 12, 24, 48, 96, ...

To **Raise your grade** now try questions 6, 9 and 19 on pages 95–96

2.8 Proportion

YOU NEED TO:

- **E** Express direct and inverse proportion in algebraic terms and use this form of expression to find unknown quantities.

Extended

$y \propto x$ means that y is directly proportional to x.

$y \propto x$ can be rewritten as $y = kx$, where k is a constant that you can find.

For example, the number of apples, A, on a tree is directly proportional to the number of branches, b. If b doubles, then A doubles.

In a similar way, $y \propto (x + 1)^2$ means that y is directly proportional to $(x + 1)^2$. This can be rewritten as $y = k(x + 1)^2$, where k is a constant.

$y \propto \dfrac{1}{x}$ means y is inversely proportional to x. This can be written as $y = \dfrac{k}{x}$, where k is a constant.

For example, the average speed of a runner, v, in a race is inversely proportional to the time, T, he takes to run the race. If T doubles, then v halves.

In a similar way, $y \propto \dfrac{1}{\sqrt{x+2}}$ means that y is inversely proportional to $\sqrt{x+2}$.

This can be written as $y = \dfrac{k}{\sqrt{x+2}}$, where k is a constant.

◀◀ RECAP

Two variables are directly proportional if there is always a constant ratio between them.

Solving problems involving proportion

In all cases, the method is the same:

- Write the relationship between x and y as an equation using k.
- Find k using a pair of values given in the question and replace k in the equation with this value.
- Use the equation to find other values of x and y.

🔑 KEY SKILLS

You must be able to express direct proportion in algebraic terms and use this form of expression to find unknown quantities.

WORKED EXAMPLE

y is inversely proportional to x and $y = 24$ when $x = 5$. Find the value of:

a) y when $x = 2$
b) x when $y = 30$. **[2 marks]**

Write the relationship between x and y as an equation using k.

$y = \dfrac{k}{x}$

Find k using the pair of values given in the question.

When $x = 5$, $y = 24$, so $24 = \dfrac{k}{5}$ and $k = 120$.

Hence $y = \dfrac{120}{x}$

🔑 KEY SKILLS

You must be able to express inverse proportion in algebraic terms and use this form of expression to find unknown quantities.

👁 WATCH OUT!

Sometimes a question may say 'x varies as y' rather than 'x is directly proportional to y'.

These mean exactly the same thing.

2 Algebra and graphs

→ Use the equation to find values of x and y.

a) When $x = 2$, $y = \dfrac{120}{2} = 60$

b) When $y = 30$, $30 = \dfrac{120}{x}$ and so $x = 4$

? QUESTIONS

1. Rewrite these statements using a constant k.
   a) $S \propto t$
   b) $F \propto \dfrac{1}{r^2}$
   c) G varies as r^2
   d) E is proportional to m

2. P is proportional to S. If $P = 16$ when $S = 4$, calculate:
   a) the value of P when $S = 7$
   b) the value of S when $P = 80$.

3. Competitors in a 'World's Strongest Man' competition are required to carry large spherical stones called Atlas Stones, which range from 100 to 160 kilograms.

   If the mass of each stone is proportional to the cube of its radius, and a 100 kg stone has a radius of 21.9 cm, what will be the radius of a 160 kg stone? (Give your answer correct to 1 decimal place.)

4. The distance, s, travelled by an object is directly proportional to the square of the time, t, for which it has been travelling. When $t = 5$, $s = 75$.
   a) Evaluate k, the constant of proportionality, and write down an equation for s in terms of t.
   b) Find the value of s when $t = 7$.
   c) Find t when $s = 363$.
   d) Describe what happens when t is doubled.

5. The mass, m, of an object is directly proportional to the cube of its side length, l.

   The mass of a cube with side length 3 cm is found to be 216 g.
   a) Calculate the constant of proportionality and write down an equation for m in terms of l.
   b) Find the mass of an object with side length 7 cm.
   c) Find the side length of an object which has mass 9261 g.
   d) Describe what happens to l when m increases by 33.1%.

6. The light intensity, l, is measured at a distance d away from a lamp.

   It is found that $l \propto \dfrac{1}{d^2}$.

   It is observed that $l = 180$ when $d = 7$.
   a) Find the constant of proportionality and write down an equation involving l and d.
   b) Find the value of l when $d = 2$.
   c) Find the value of d when $l = 45$.
   d) Describe what happens to d when l decreases by 75%.

7. The Eiffel Tower in Paris is repainted every 7 years.

   It takes 25 painters 18 months to complete the repainting.

   Assuming they didn't get in each other's way, find how long the job would take:
   a) 50 painters
   b) 75 painters.

   To **Raise your grade** now try question 22 on page 97

2.9 Real-life graphs

> **YOU NEED TO:**
> - Use and interpret graphs in practical situations including travel graphs and conversion graphs.
> - Draw graphs from given data.
> - **Ⓔ** Apply the idea of rate of change to simple kinematics involving distance–time and speed–time graphs, acceleration and deceleration.
> - **Ⓔ** Calculate distance travelled as area under a speed–time graph.

RECAP

Graphs are often useful for practical things, such as performing unit conversions (for example, converting one currency into another) or analysing a journey.

KEY SKILLS

You must be able to interpret and use conversion graphs.

WORKED EXAMPLE

The graph shows the amount that a shop charges for hiring a bike for up to 8 hours in a day.

There is an initial charge and then an hourly charge.

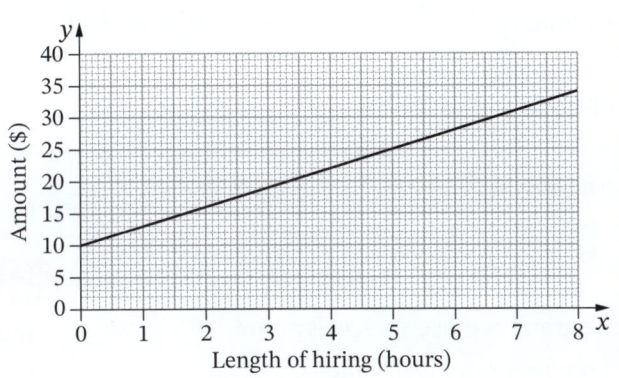

a) What is the initial charge? **[1 mark]**

b) What is the hourly charge? **[1 mark]**

c) How much would it cost to hire a bike for 3 hours? **[1 mark]**

d) How many hours' hire would cost $28? **[1 mark]**

a) The initial charge is the cost when the time is zero. This is $10.

b) From the graph you can see that 1 hour costs $13. So the hourly charge is $3.

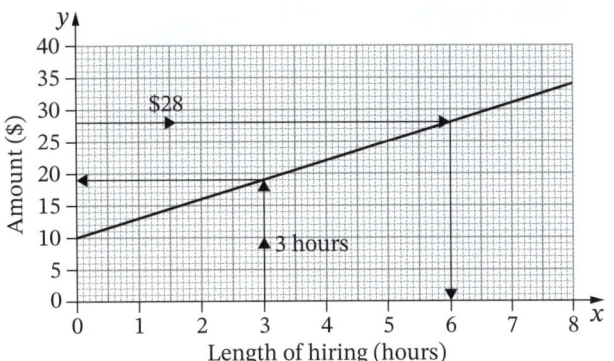

c) Reading off from the graph gives $19.

d) Reading off from the graph gives 6 hours.

79

2 Algebra and graphs

Extended

Distance–time graphs

Gradient on a distance–time graph represents speed.

This graph shows the journey of a cyclist. From the graph you can tell how the cyclist's speed changes.

What is happening to the speed…

Between 0 s and 2 s the cyclist is speeding up.

Between 2 s and 5 s the cyclist is travelling at a constant speed.

Between 5 s and 7 s the cyclist is slowing down.

Between 7 s and 10 s the cyclist is stationary.

Between 10 s and 12 s the cyclist is speeding up.

From 12 s to 15 s the cyclist is travelling at a constant speed.

From 15 s to 17 s the cyclist is slowing down.

> **KEY SKILLS**
> You must be able to interpret travel graphs.

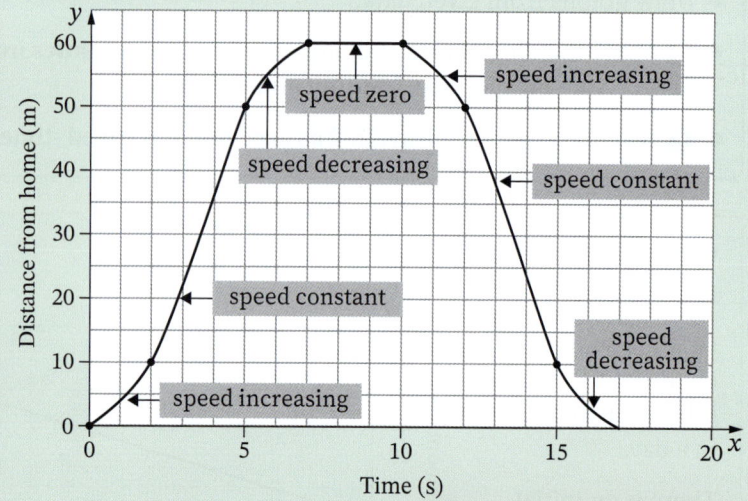

Speed–time graphs

Gradient on a speed–time graph represents acceleration.
Area under a speed–time graph represents distance travelled.

This graph shows the speed of a toy car. From the graph you can tell how the car's speed and acceleration change.

What is happening to the speed

Between 0 s and 17 s the car is speeding up to reach a top speed of $12\,m\,s^{-1}$.

Between 17 s and 25 s the car has a constant speed of $12\,m\,s^{-1}$.

From 25 s onwards the car is slowing down (decelerating) to come to rest.

What is happening to the acceleration

Between 0 s and 2 s the acceleration is increasing.

Between 2 s and 12 s the acceleration is constant.

Between 12 s and 17 s the acceleration is decreasing.

Between 17 s and 25 s the acceleration is zero.

Between 25 s and 30 s the deceleration is increasing.

From 30 s to 35 s the deceleration is constant.

From 35 s to 40 s the deceleration is decreasing.

> **KEY SKILLS**
> You must be able to apply the idea of rate of change to simple kinematics problems involving distance–time and speed–time graphs, and acceleration and deceleration.

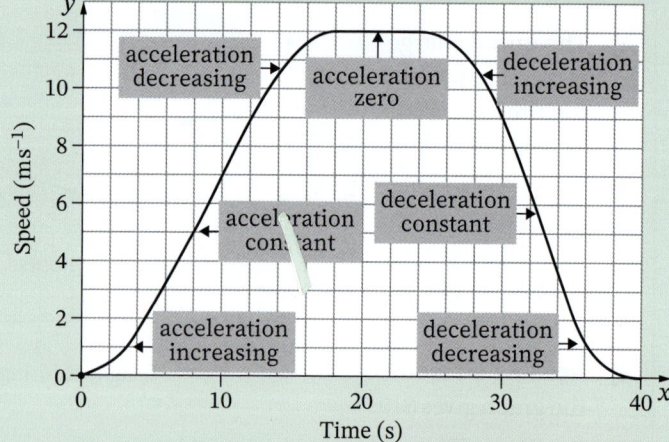

> **KEY SKILLS**
> You must be able to calculate distance travelled as area under a speed–time graph.

2.9 Real-life graphs

WORKED EXAMPLE

A coach was hired to take a school team to a football game. The graph shows the distance travelled from the school.

a) How far did the coach travel before stopping? **[1 mark]**

b) For how long did the coach stop? **[1 mark]**

c) How much further did the coach travel to the game? **[1 mark]**

d) How long did the coach wait at the game? **[1 mark]**

e) How long did the return journey take? **[1 mark]**

f) What was the average speed of the coach on the return journey? **[3 marks]**

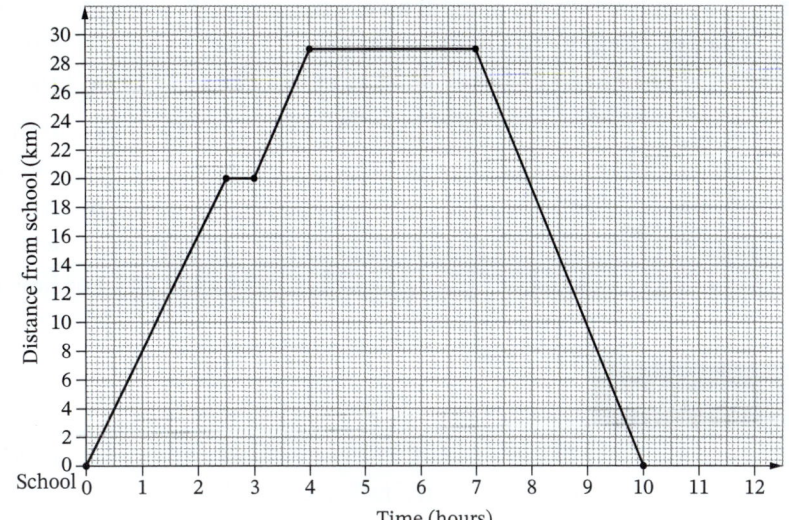

a) The first line segment stops at a height of 20 km, so the answer is 20 km.

b) The first flat line segment goes from 2.5 to 3 on the horizontal axis, so the answer is 30 min.

c) The next time the coach stops is at 29 km. 29 − 20 = 9, so the answer is another 9 km.

d) The second flat line segment goes from 4 to 7 on the horizontal axis, so the answer is 3 hours.

e) The final section of the graph goes from 7 to 10 on the horizontal axis, so the answer is 3 hours.

f) On the return journey, they travel 29 km in 3 hours, so the average speed is $29 \div 3 = 9.67 \, \text{km}\,\text{h}^{-1}$.

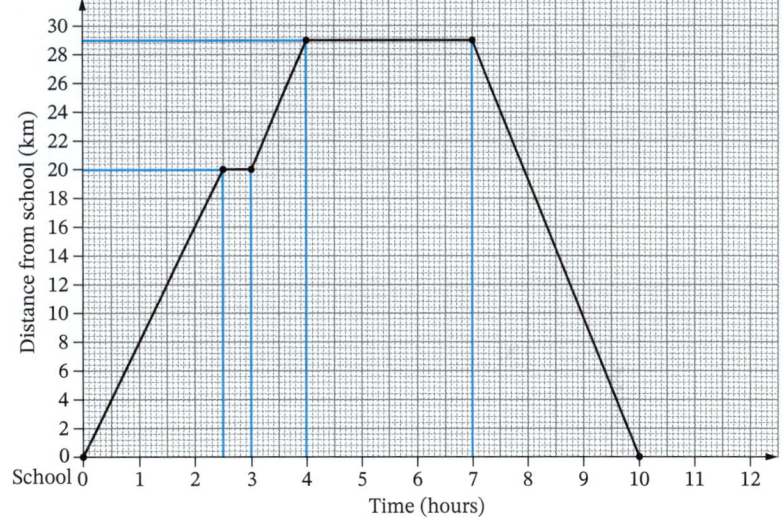

EXAM TIP

Begin by drawing horizontal and vertical lines that separate the graph at the important places, and show clearly where these places are, relative to the horizontal and vertical axes.

EXAM TIP

Remember: speed = distance ÷ time

In the following worked example, distance is in kilometres and time is in hours, so speed is in kilometres per hour.

'kilometres' are a 'distance', 'per' is 'divided by' and 'hour' is a 'time'.

This idea can be used with different units of compound measure.

2 Algebra and graphs

? QUESTIONS

1. The graph shows the amount an electrician charges for up to five hours' work.

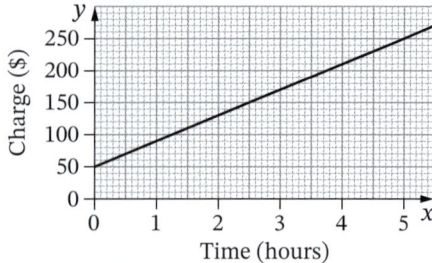

a) What does the electrician charge for being called out?

b) How much does he charge for 2 hours' work?

c) If the electrician charged Mr Bali $210, how many hours work did they do?

d) A second electrician has a call out fee of $60 and an hourly fee of $35.

For how many hours' work would the two electricians charge the same amount?

Extended

2. Hasnain drove to see a friend. The graph shows his journey to and from his friend's house and the time he spent with his friend.

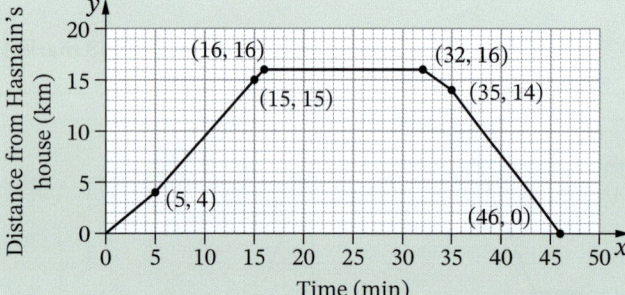

a) How long did it take for Hasnain to reach a constant speed?

b) What was happening to his speed in the first 5 minutes?

c) What was the constant speed in $km\,h^{-1}$?

d) How long did Hasnain spend at his friend's house?

e) What was his constant speed for the last 11 minutes of his journey?

3. The graph shows information about a journey that Awais made by car.

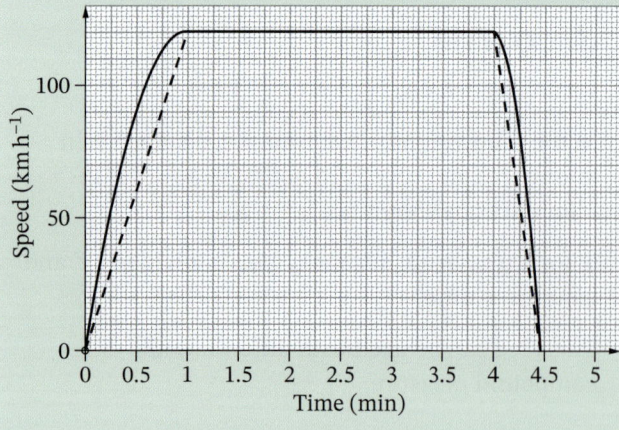

a) How long did it take Awais to reach a constant speed?

b) How far did he travel at this constant speed?

c) What was happening to his acceleration in the first minute?

d) By using the dotted lines for the first minute and the last 30 seconds, estimate how far Awais travelled.

e) Is your answer to part d) an under-estimate or an over-estimate of the distance that Awais travelled? Give a reason for your answer.

4. The graph shows the journey made by Karen on her bicycle. She cycled from her home to a café 50 km away and then back.

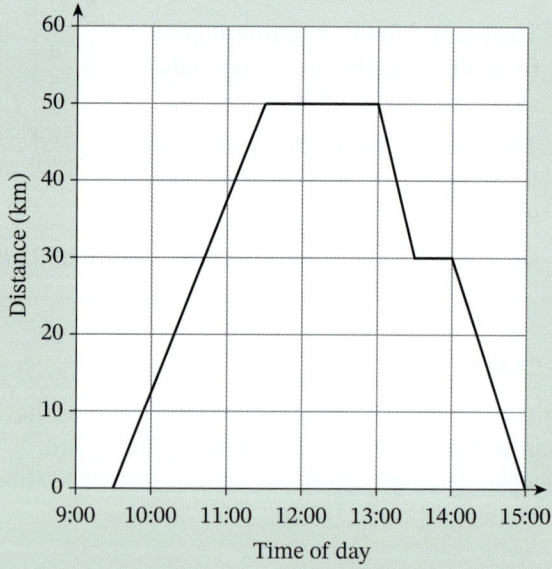

a) How long did she stay at the café?

b) What was her speed during the first part of her journey?

c) What was her average speed while cycling on the way back?

To **Raise your grade** now try question 20 on page 97

2.10,11 Understanding graphs

YOU NEED TO:

- Construct tables of values, and draw, recognise and interpret graphs for functions of the following forms: $ax + b$, $\pm x^2 + ax + b$, $\frac{a}{x} (x \neq 0)$, where a and b are integer coefficients.
- **E** Construct tables of values, and draw, recognise and interpret graphs for functions of ax^n (includes sums of no more than three of these) and $ab^x + c$, where $n = -2, -1, -\frac{1}{2}, 0, \frac{1}{2}, 1, 2, 3$; a and c are rational numbers; and b is a positive integer.
- Solve associated equations graphically, including finding and interpreting roots by graphical methods.
- **E** Draw and interpret graphs representing exponential growth and decay problems.
- Recognise, sketch and interpret graphs of linear and quadratic functions.
- **E** Recognise, sketch and interpret graphs of cubic, reciprocal and exponential functions.

WORKED EXAMPLE

The table gives the values of x and y for the function $y = 3x + 2$.

x	−1	0	1	2	3	4	5
y		2	5	8	11		

a) Fill in the missing values of y. [1 mark]
b) Sketch the graph of $y = 3x + 2$. [2 marks]
c) Use your graph to solve the equation $3x + 2 = 10$. [1 mark]

a) To find the y-value when $x = -1$ substitute -1 for x into $3x + 2$.
$3 \times -1 + 2 = -1$
Repeating this for $x = 4$ and $x = 5$ gives:

x	−1	0	1	2	3	4	5
y	−1	2	5	8	11	14	17

b) Mark the points on the graph, that is, (−1, −1), (0, 2), etc.

Join the points using a straight line.

KEY SKILLS
You need to be familiar with several different types of graphs.

KEY SKILLS
You must be able to construct tables of values for functions of the form $ax + b$, $\pm x^2 + ax + b$, $\frac{a}{x} (x \neq 0)$, where a and b are integers, draw and interpret these graphs, and use them to find approximations to roots of equations.

Extended

KEY SKILLS
The graphs may also include functions of the form ax^n (and simple sums of these) and functions of the form $ab^x + c$.

2 Algebra and graphs

c) The solution to $3x + 2 = 10$ is the x-coordinate of the point on the graph where $y = 3x + 2$ meets $y = 10$.

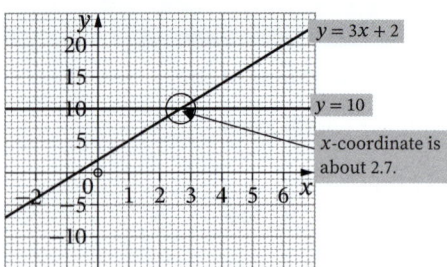

x-coordinate is about 2.7.

From the graph, the solution to $3x + 2 = 10$ is 2.7 (to 1 d.p.).

> **EXAM TIP**
>
> When plotting the line, you don't have to plot *all* the points, but you should plot at least 3 of them. Remember that the x-axis is horizontal and that the y-axis is vertical (alphabetical order in both): $x, y; h, v$.

WORKED EXAMPLE

The table gives the values of x and y for the function $y = x^2 - 4x + 4$.

x	−1	0	1	2	3	4	5
y		4	1		1		9

a) Fill in the missing values of y. **[1 mark]**
b) Sketch the graph of $y = x^2 - 4x + 4$. **[2 marks]**
c) Use the graph to solve $x^2 - 4x + 4 = 7$. **[2 marks]**

a) To find the value of y when $x = -1$, substitute −1 for x into $x^2 - 4x + 4$.
Use brackets on the calculator, so type in $(-1)^2 - 4(-1) + 4$ to get 9.
Repeating this for the values 2 and 4 gives:

x	−1	0	1	2	3	4	5
y	9	4	1	0	1	4	9

b) Mark the points on the graph, that is, (−1, 9), (0, 4), etc.

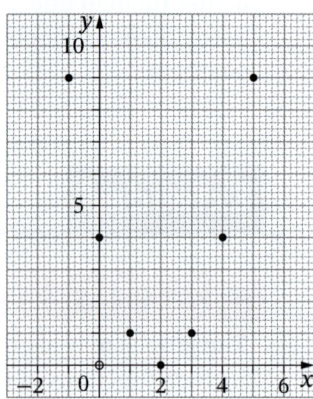

Join the points using a smooth curve.

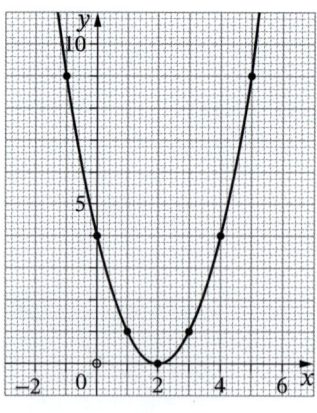

> **EXAM TIP**
>
> The basic shape of the graph of $y = ax^2 + bx + c$ is shown in these diagrams.
>
> i) $a > 0$
>
>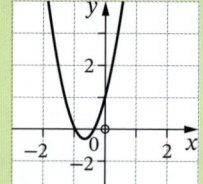
>
> ii) $a < 0$
>
>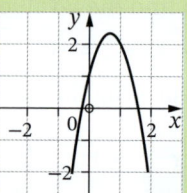
>
> A positive value of a gives the graph a smiley shape, because it is feeling positive.
>
> A negative value of a gives the graph a frowny shape, because it is feeling negative.

c) The solutions to $x^2 - 4x + 4 = 7$ are the x-coordinates of the points on the graph where $y = x^2 - 4x + 4$ meets $y = 7$.

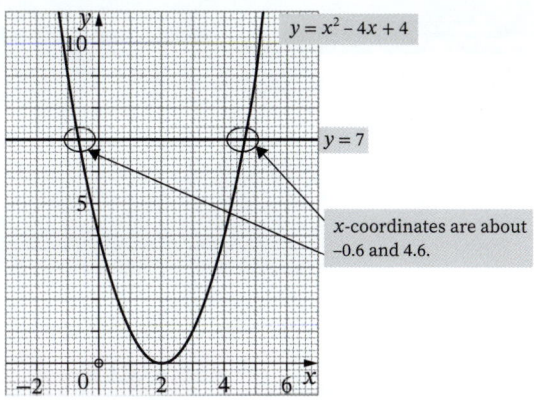

From the graph, the solutions to $x^2 - 4x + 4 = 7$ are −0.6 and 4.6 (to 1 d.p.).

WORKED EXAMPLE

The table gives the values of x and y for the function $y = \dfrac{24}{x}$.

x	−4	−3	−2	−1	0	1	2	3	4
y	−6	−8			−		12	8	

a) Fill in the missing values of y. [1 mark]

b) Sketch the graph of $y = \dfrac{24}{x}$. [2 marks]

a) To find the value of y when $x = -2$, substitute −2 for x into $\dfrac{24}{x}$ to give $y = -12$.

Repeating this for −1, 2 and 4 gives:

x	−4	−3	−2	−1	0	1	2	3	4
y	−6	−8	−12	−24	−	24	12	8	6

b) Mark the points on the graph, that is, (−4, −6), (−3, −8) etc. Join up the points using a smooth curve.

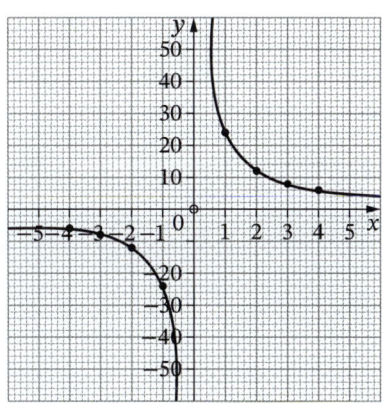

WATCH OUT!

Notice that in the table there is no y value for when $x = 0$. This is because you cannot divide by zero.

APPLY

Choose any number and, on your calculator, divide it by 0.

You should always get an error, no matter what number you choose.

2 Algebra and graphs

Extended

◀◀ RECAP

An exponential graph is a graph of a function where the variable (usually x) is the power, rather than the base.

For example:

x^2 is a quadratic function.

2^x is an exponential function.

Graph of $y = a^x$

When $a > 0$ the graph of $y = a^x$ has this basic shape:

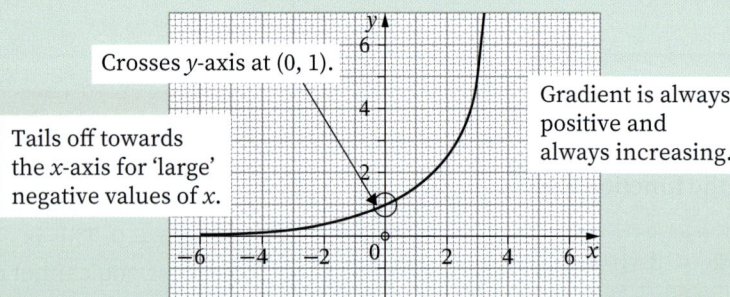

Crosses y-axis at (0, 1).

Gradient is always positive and always increasing.

Tails off towards the x-axis for 'large' negative values of x.

🔑 KEY SKILLS

You must be able to draw and interpret graphs representing exponential growth and decay problems.

👍 EXAM TIP

Remember that exponential graphs all have this basic shape. You deal with them in exactly the same way as you would any other graph.

Any function of the form $y = a^x$ will cross the y-axis at (0, 1).

Graph of $y = ax^{-2}$

When the power of x is –2, the graph is based on the reciprocal of x^2.

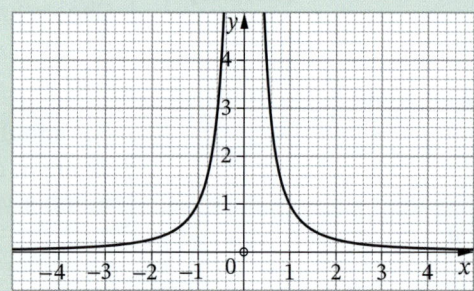

The right-hand branch is similar to the right-hand branch of $y = \dfrac{1}{x}$, but the left-hand branch is also positive because x is squared.

Graph of $y = ax^{-\frac{1}{2}}$

When the power of x is $-\dfrac{1}{2}$, the graph will have a single branch. Both axes will be asymptotes.

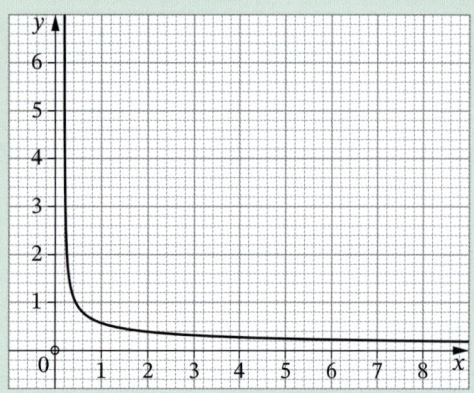

Graph of $y = ax^{\frac{1}{2}}$

When the power of x is $\frac{1}{2}$, the graph is based on the square root of x. Remember this function is only defined for positive values of x.

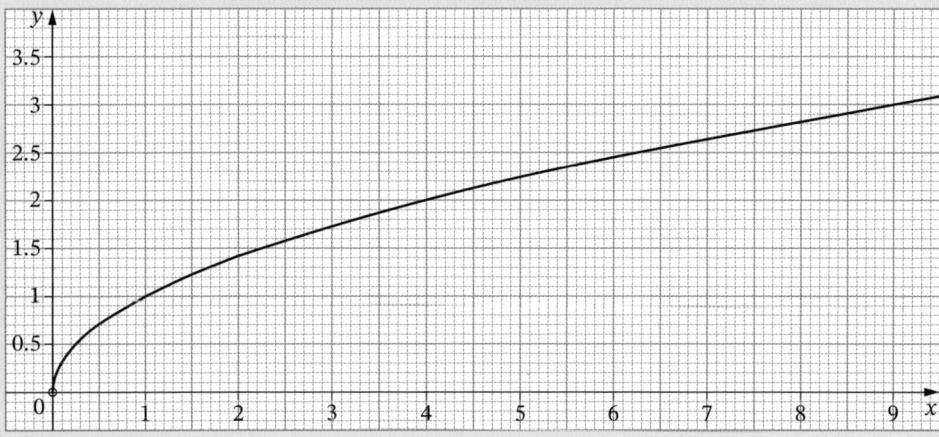

Graph of $y = ax^3$

When the power of x is 3, the graph will look like a steeper version of the quadratic graph, except that the left-hand half of the graph is under the x-axis.

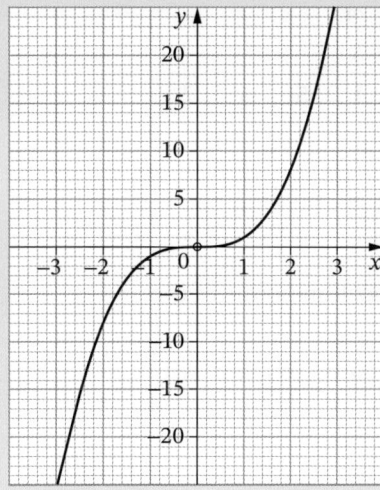

2 Algebra and graphs

QUESTIONS

1. This table of values is for the graph $y = x^2 - 6x + 1$.

x	0	1	2	3	4	5	6
y		−7					

   a) Copy and complete the table.
   b) Draw a scale from 0 to 6 on the x-axis (2 cm per unit) and from −10 to 5 on the y-axis (1 cm per unit).
   c) Draw the graph of $y = x^2 - 6x + 1$.
   d) Use your graph to find the values of x (to 1 d.p.) when $y = 0$ and $y = -5$.

2. This table of values is for $y = x^2 - 2x - 20$.

x	−5	−4	−3	−2	−1	0	1	2	3	4	5	6	7
y	15	4	−5		−17			−20	−17		−5	4	

   a) Copy and complete the table.
   b) Draw a scale from −5 to 7 on the x-axis (1 cm per unit) and from −30 to 40 on the y-axis (2 cm per 5 units). Draw the graph of $y = x^2 - 2x - 20$.
   c) Copy and complete the table of values for $y = \dfrac{24}{x}$.

x	−5	−4	−3	−2	−1	0	1	2	3	4	5	6	7
y	−4.8	−6.0			−24.0	−	24.0	12.0	8.0		4.8		3.4

   d) On the same diagram as part b), draw the graph of $y = \dfrac{24}{x}$ for $-5 \leq x \leq 7$.
   e) Write down the x-coordinate of the two intersection points of the two graphs.
   These two x values are the solutions to an equation. Write down and simplify this equation.

3. The table of values is for the graph $y = x^2 - 2x - 4$.

x	−3	−2	−1	0	1	2	3	4	5
y							4		

   a) Copy and complete the table.
   b) Draw a scale from −3 to 5 on the x-axis (2 cm per unit) and from −8 to 12 on the y-axis (1 cm per unit).
   c) Draw the graph of $y = x^2 - 2x - 4$.
   d) Use your graph to solve (to 1 d.p.) the equation $x^2 - 2x - 4 = 0$.
   e) Use your graph again to solve (to 1 d.p.) the equation $x^2 - 2x - 4 = 5$.
   f) What is the smallest value of $x^2 - 2x - 4$ and which value of x achieves this smallest value?

Extended

4. Draw the graph of $y = 2^x$ using x values from −1 to 5 (1 cm per unit) and a scale of 1 cm per 5 units on the y-axis. Find the approximate value of x when $y = 10$.

2.12 Differentiation

> **YOU NEED TO:**
> - **E** Estimate gradients of curves by drawing tangents.
> - **E** Use the derivatives of functions of the form ax^n, where a is a rational constant and n is a positive integer or zero, and simple sums of not more than three of these.
> - **E** Apply differentiation to gradients and stationary points (turning points).
> - **E** Discriminate between maxima and minima by any method.

Extended

> **◀◀ RECAP**
>
> Unlike straight lines, finding the gradient of a curve at a given point can be tricky.
>
> One way to estimate it is to draw a tangent to the curve and find the gradient of the tangent.

WORKED EXAMPLE

Approximate the gradient of the curve $y = x^2 + 1$ at the point (2, 6) by drawing a tangent. **[3 marks]**

First plot the graph as accurately as you can, near to the point where you want to find the gradient.

Draw a tangent and then, using the most convenient points available, make a right-angled triangle underneath it. From the sides of the triangle, you can work out the gradient of the line.

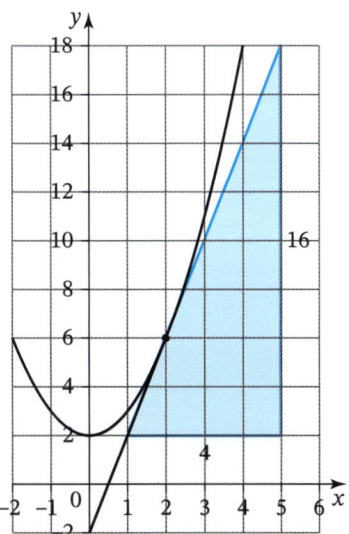

In this example, the estimate of the gradient is $16 \div 4 = 4$.

> 🔑 **KEY SKILLS**
>
> You must be able to estimate gradients of curves by drawing tangents.

> 👍 **EXAM TIP**
>
> Remember that the gradient of a curve at a point is the same as the gradient of the tangent to the curve at that point.

2 Algebra and graphs

> ⏪ **RECAP**
>
> A derived function is obtained by differentiating another function.
>
> Differentiation is part of an important area of mathematics called calculus.

> 👁 **WATCH OUT!**
>
> Be careful to check how the axes are calibrated. Don't just count the squares.

Given any function $y = f(x)$, the function $\frac{dy}{dx}$ is obtained by differentiating the original function.

Although differentiation has many applications, you should primarily think of $\frac{dy}{dx}$ as being the function that tells you the gradient of $y = f(x)$ for any given value of x.

> ⏪ **RECAP**
>
> The derivative of a function of the form ax^n is anx^{n-1}.
>
> In other words, multiply by the power, then subtract 1 from the power.
>
> $\frac{d}{dx} ax^n = anx^{n-1}$

> 🗝 **KEY SKILLS**
>
> You must understand the idea of a derived function.

> ⏪ **RECAP**
>
> If the expression has several terms, you just differentiate each term, one at a time.

> 🗝 **KEY SKILLS**
>
> You must be able to use the derivatives of functions of the form ax^n, and simple sums of not more than three of these.

> 👁 **WATCH OUT!**
>
> It is not a good idea, when you are new to calculus, to think of $\frac{dy}{dx}$ as being a fraction. It is better just to think of it as being the name of a function.

> 👍 **EXAM TIP**
>
> The derived function is often called the derivative. They mean exactly the same thing.

WORKED EXAMPLE

Differentiate the following functions.

a) $y = x^2$ [1 mark]
b) $y = 3x^4$ [1 mark]
c) $y = 5x^2 + 3x + 2$ [1 mark]

a) $\frac{dy}{dx} = 2 \times x = 2x$

b) $\frac{dy}{dx} = 4 \times 3x^{4-1} = 12x^3$

c) $\frac{dy}{dx} = 2 \times 5x^{2-1} + 3 = 10x + 3$

> 👍 **EXAM TIP**
>
> $\frac{d}{dx}$ is a function that means 'the derivative of'.

> 👍 **EXAM TIP**
>
> Remember that when you differentiate a constant term, it becomes zero (i.e. it disappears).

2.12 Differentiation

⏪ RECAP

A stationary point on a graph is any place where the gradient is zero.

A turning point is a type of stationary point.

A turning point can be a 'local maximum' or a 'local minimum'.

Usually you just say 'maximum' or 'minimum', and if you are talking about more than one you call them 'maxima' or 'minima'.

🔑 KEY SKILLS

You must be able to apply differentiation to gradients and turning points (stationary points).

WORKED EXAMPLE

Find the x-coordinates of any stationary points of the function $y = x^2 - 6x$. **[3 marks]**

If $y = x^2 - 6x$, then $\frac{dy}{dx} = 2x - 6$

A stationary point is where $\frac{dy}{dx} = 0$, so $2x - 6 = 0$

Therefore $x = 3$.

👍 EXAM TIP

Note that this turning point could probably have been found more quickly by completing the square.

✏️ APPLY

Find the turning point of the quadratic curve in the above Worked example by completing the square.

⏪ RECAP

You need to be familiar with two methods for discriminating between different types of turning points.

1. Using the second derivative.
2. Considering the gradients on both sides of the point.

🔑 KEY SKILLS

You must be able to discriminate between maxima and minima by any method.

👍 EXAM TIP

To 'discriminate' means to recognise which kind of turning point it is.

WORKED EXAMPLE

Find the x-coordinates of the turning points of the function $y = \frac{1}{3}x^3 - \frac{1}{2}x^2 - 2x - 1$ and determine the nature of the points. **[5 marks]**

If $y = \frac{1}{3}x^3 - \frac{1}{2}x^2 - 2x - 1$, then $\frac{dy}{dx} = x^2 - x - 2$

A stationary point is where $\frac{dy}{dx} = 0$, so $x^2 - x - 2 = 0$

This factorises as $(x - 2)(x + 1) = 0$

Therefore $x = -1$ or $x = 2$.

Method 1 – Using the second derivative

$\frac{d^2y}{dx^2} = 2x - 1$

When $x = -1$, $\frac{d^2y}{dx^2} = (2 \times -1) - 1 = -3$, which is negative, so $x = -1$ is a maximum.

👍 EXAM TIP

To 'determine their nature' is another phrase commonly used to mean 'work out which kind of turning point they are'.

👍 EXAM TIP

$\frac{d^2y}{dx^2}$ is called 'the second derivative'.

To find $\frac{d^2y}{dx^2}$ you differentiate the expression for $\frac{dy}{dx}$.

2 Algebra and graphs

When $x = 2$, $\frac{d^2y}{dx^2} = (2 \times 2) - 1 = 3$, which is positive, so $x = 2$ is a minimum.

Method 2 – Considering the gradients on both sides of the point

When $x = -1$, you need to look at the gradients on each side.

x	−2	−1	0
$\frac{dy}{dx}$	4	0	−2

Because the gradient is positive before the stationary point and negative directly after it, you can tell that $x = -1$ is a maximum.

When $x = 2$, you need to look at the gradients on each side.

x	1	2	3
$\frac{dy}{dx}$	−2	0	4

Because the gradient is negative before the stationary point and positive directly after it, you can tell that $x = 2$ is a minimum.

> **EXAM TIP**
> Remember that if the second derivative is also zero, you must use method 2.

> **WATCH OUT!**
> Sometimes it might seem obvious which point is a maximum and which is a minimum, because you know the shape of the graph, but you still need to prove it using algebra.

> **WATCH OUT!**
> By taking a step of one whole unit either side of the stationary point, you must be careful that nothing else significant happens to the curve within that interval.
>
> Exam questions, however, are usually designed so that nothing significant does happen.
>
> You can, if you prefer, use points that are nearer to the stationary point.

QUESTIONS

1. Draw a tangent to find an approximation of the gradient of the curve $y = (x - 2)^2 + 1$ when $x = 3$.

2. Draw a tangent to find an approximation of the gradient of the curve $y = x^2 - 3x + 5$ when $x = 2$.

3. Draw a tangent to find an approximation of the gradient of the curve $y = 5 - x^2$ at the point $(2, 1)$.

4. Find the coordinates of the stationary point of the function $y = 2x^2 + 4x - 3$ and determine its nature.

5. Find the turning point of the function $y = x^4 + 1$ and determine its nature.

6. Find the x-coordinates of the stationary points of the function $y = x^3 - 3x^2 - 24x + 8$ and determine their nature.

To **Raise your grade** now try questions 15, 17 and 24 on pages 96–97

2.13 Functions

YOU NEED TO:
- **E** Understand functions, domain and range and use function notation.
- **E** Understand and find inverse functions $f^{-1}(x)$.
- **E** Form composite functions as defined by $gf(x) = g(f(x))$.

Extended

◀◀ RECAP

A function is a mapping in which each element in the domain has one, and only one, image in the range.

Functions are usually denoted by the letters f, g, etc. So you can write

$f : x \mapsto x + 1$ or $f(x) = x + 1$ and $g : x \mapsto x^2$ or $g(x) = x^2$

🗝 KEY SKILLS

You must be able to use function notation, e.g. $f(x) = 3x - 5$, $f : x \mapsto 3x - 5$, to describe simple functions.

◀◀ RECAP

To find the inverse of $f(x)$:
- put $f(x) = y$ and make x the subject of the formula in terms of y
- replace each y with x to find $f^{-1}(x)$.

👍 EXAM TIP

The $f(x)$ notation and the arrow notation are used interchangeably. They mean exactly the same thing.

WORKED EXAMPLE

Find the inverse of $f(x) = \dfrac{3x - 5}{2}$ [3 marks]

Putting $f(x) = y$: $y = \dfrac{3x - 5}{2}$

Making x the subject of the formula: $x = \dfrac{2y + 5}{3}$

Replacing y with x gives $f^{-1}(x) = \dfrac{2x + 5}{3}$

🗝 KEY SKILLS

You must be able to find inverse functions $f^{-1}(x)$.

👁 WATCH OUT!

Only one-to-one functions have inverses.

◀◀ RECAP

If $f(x) = x + 1$ and $g(x) = x^2$ then 'gf' means 'do f first, then g'.

So $gf(x)$ means $g(f(x))$.

🗝 KEY SKILLS

You must be able to form composite functions as defined by $gf(x) = g(f(x))$.

2 Algebra and graphs

WORKED EXAMPLE

If $f(x) = x + 3$ and $g(x) = x^2$ then find:

a) the value of x for which $f(x) = 8$ [2 marks]
b) the values of x for which $g(x) = 36$ [2 marks]
c) $fg(4)$ [2 marks]
d) $gf(x)$ [2 marks]

a) Solving $f(x) = 8$ means solving $x + 3 = 8$. So $x = 5$.
b) Solving $g(x) = 36$ means solving $x^2 = 36$. So $x = 6$ or $x = -6$.
c) $g(4) = 4^2 = 16$ so $f(g(4)) = f(16) = 16 + 3 = 19$
d) $f(x) = x + 3$ so $gf(x) = g(x + 3) = (x + 3)^2$

> **WATCH OUT!**
>
> $fg(x)$ is not the same as $gf(x)$.
>
> Work from the inside outwards.

? QUESTIONS

1. If $f(x) = x^2 + 3$, find:
   a) $f(2)$
   b) $f(-1)$
   c) a value of x such that $f(x) = 3$.

2. Given the function $f: x \mapsto \dfrac{3x^2 + 1}{2}$ find:
   a) $f(0)$
   b) $f(2)$
   c) $f(-3)$

3. If $f(x) = \dfrac{x + 1}{x - 2}$, find $f^{-1}(x)$

4. If $f(x) = x^2$ and $g(x) = x + 1$, find:
   a) $fg(2)$
   b) $fg(x)$
   c) a value of x such that $fg(x) = 16$
   d) $gg^{-1}(-3)$

5. The functions f and g are as follows:
   $f: x \mapsto 2x + 5$
   $g: x \mapsto 2 + \sqrt{x}$
   a) Calculate $f(-3)$.
   b) Given that $f(a) = 17$, find the value of a.
   c) Find the inverse function of g.

6. Find the inverses of the following functions, in the form '$x \mapsto$ ...'
   a) $f: x \mapsto 4x - 1$
   b) $f: x \mapsto \dfrac{3(x + 4)}{2}$

> To **Raise your grade** now try questions 18, 21 and 23 on pages 96–97

↑ Raise your grade

1. Simplify $\dfrac{6x^4 y^2}{3x^2 y}$ [2 marks]

2. Find the point of intersection of these straight lines:
   $3x + 4y = 22$
   $-5x + 2y = 24$ [4 marks]

3. The volume V of a cone with base radius r and height h is given by the formula $V = \dfrac{1}{3}\pi r^2 h$
   a) Calculate the volume of a cone with base radius 6 cm and height 12 cm. [2 marks]
   b) Rearrange the formula to make h the subject. [1 mark]
   c) A cone has a volume of 960 cm³ and a base radius of 12 cm.
      Calculate its height. [2 marks]
   d) A cone has a volume of 72π cm³, and its radius and height are equal.
      Calculate its height. [2 marks]

4. Find the value of x in this equation: $\dfrac{7}{5} + \dfrac{16}{10} = \dfrac{x}{4}$ [3 marks]

5. Solve the inequality $\dfrac{x-3}{5} < \dfrac{2x-3}{4}$ [3 marks]

6. A sequence begins 10, 7, 4, 1, −2, ...
   Write down:
   a) the nth term rule [1 mark]
   b) the smallest value of n for which the nth term is less than −100. [1 mark]

7. The diameter of Saturn is 10 times the diameter of Venus, and the diameter of Venus is 2.4 times the diameter of Neptune. If the diameter of Saturn is x times the diameter of Neptune, what is x? [5 marks]

8. If $x = -\dfrac{1}{4}$, from the expressions x^{-2}, x^{-1}, x^0, x, and x^2, write down the one that has:
   a) the largest value [3 marks]
   b) the smallest value. [1 mark]

Extended

9. A sequence begins 1, 3, 9, 27, 81,...
   Write down:
   a) the nth term rule [1 mark]
   b) the 15th term in the sequence. [1 mark]

10. Expand and simplify $(x + 1)(2x + 1)(x - 3)$. [2 marks]

11. In this diagram, the rectangle $ABCD$ is attached to the top of the trapezium $CDEF$. $AB = 2(x + 1)$ cm, $BC = 2(x - 1)$ cm, $EF = 4x$ cm, and the perpendicular height of the trapezium is $(2x - 3)$ cm.

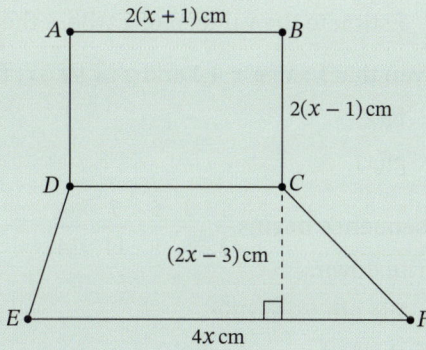

a) Write an expression for the area of rectangle ABCD. [1 mark]
b) Write an expression for the area of trapezium CDEF. [1 mark]
c) Write an expression for the total area of ABCD and CDEF. [1 mark]
d) The total area of ABCD and CDEF is 91 cm². Calculate the value of x. [4 marks]

12. $x^2 - 6x + 14$ can be written in the form $(x + p)^2 + q$. Find the values of p and q. [2 marks]

13. The sum S of the first n positive integers can be found using the formula: $S = \dfrac{n(n+1)}{2}$

    If the sum of the first n positive integers is 153:
    a) show that $n^2 + n - 306 = 0$ [2 marks]
    b) solve this quadratic equation to find the value of n. [3 marks]

14. Show that the expression $\dfrac{a}{b} + \dfrac{a-b}{ab} - \dfrac{b}{a}$ can be rearranged to $\dfrac{(a-b)(a+b+1)}{ab}$ [4 marks]

15. Given that $y = 3x^3 - 4x^2 + 2x + 7$, find:
    a) $\dfrac{dy}{dx}$ [2 marks]
    b) $\dfrac{d^2y}{dx^2}$ [2 marks]

16. Factorise $3x^2 + x - 2$. [2 marks]

17. This is an accurate graph of $y = f(x)$.
    a) Use the graph to solve:
       i) $f(x) = 0$ [1 mark]
       ii) $f(x) = -1$. [1 mark]

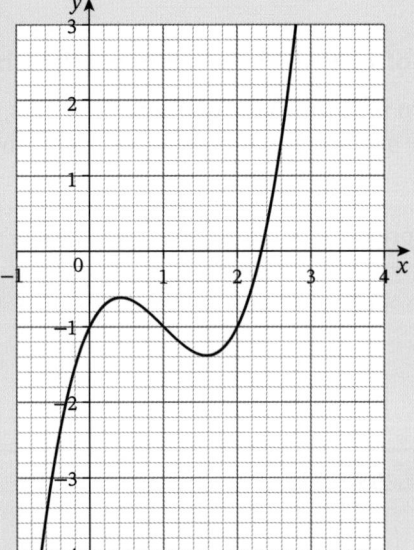

b) Write down the equation of the line that would need to be added to the graph to solve the equation $f(x) - 2x + 1 = 0$. [1 mark]
c) Estimate the gradient of the curve at the point where $x = 2$. [1 mark]
d) Estimate the value(s) of x where the gradient $= 0$. [2 marks]

18. Given that $f : x \mapsto x + 1$ and $g : x \mapsto 3x$, find the following composites in the form '$x \mapsto ...$':
    a) $fg(x)$ [2 marks]
    b) $gf(x)$ [2 marks]

19. A sequence begins $\dfrac{1}{2}, \dfrac{3}{5}, \dfrac{5}{8}, \dfrac{7}{11}, \dfrac{9}{14}, ...$
    Write down:
    a) the nth term rule [2 marks]
    b) the 20th term in the sequence. [1 mark]

20. The graph shows the train journey between Birmingham and Stratford-upon-Avon.

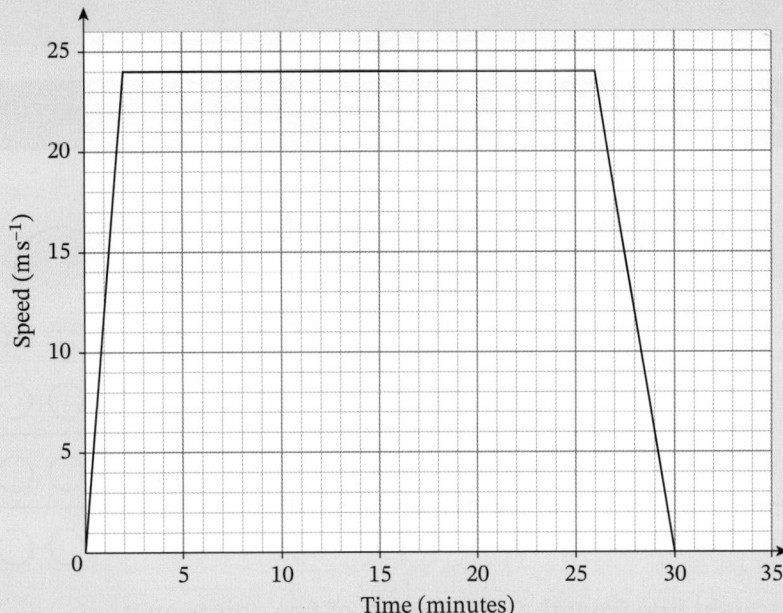

a) Find the acceleration of the train for the first two minutes of the journey. [2 marks]
b) Find the distance, in km, between Birmingham and Stratford-upon-Avon. [2 marks]
c) Find the average speed of the train journey. Give your answer in $km\,h^{-1}$. [2 marks]

21. $f(x) = x^2 + x - 2$ and $g(x) = g(x) = \dfrac{x-1}{2}$

Find:
a) $g^{-1}(x)$ [2 marks]
b) $gg^{-1}(4)$ [2 marks]
c) $fg(x)$ in the form $\dfrac{x^2 - a}{b}$, where a and b are both integers. [2 marks]

22. When a beach ball is inflated, its surface area A is proportional to the square of its radius r.
a) What happens to A when r is doubled? [2 marks]
b) What happens to A when r is decreased by 20%? [2 marks]

23. Find the inverses of the following functions, in the form '$x \mapsto ...$':
a) $f : x \mapsto \dfrac{1}{2}(x^2 + 6) + 5$ [2 marks]
b) $f : x \mapsto \dfrac{\sqrt{x^2 + 7}}{3}$ [2 marks]

24. Find the turning points of the function $f(x) = 2x^3 + 3x^2 - 36x + 4$ and determine their nature. [10 marks]

3 Coordinate geometry

Your revision checklist

Tick these circles to build a record of your revision.

Core/ **E** Extended syllabus

		😟 😐 🙂
3.1	Use and interpret Cartesian coordinates in two dimensions.	○ ○ ○
3.2	Draw straight-line graphs for linear equations.	○ ○ ○
3.3	Find the gradient of a straight line with the aid of a grid.	○ ○ ○
	E Calculate the gradient of a straight line from the coordinates of two points on it.	○ ○ ○
3.4	**E** Calculate the length of a line segment.	○ ○ ○
	E Find the coordinates of the midpoint of a line segment.	○ ○ ○
3.5	Interpret and obtain the equation of a straight-line graph in the form $y = mx + c$.	○ ○ ○
3.6	Find the gradient and equation of a straight line parallel to a given line.	○ ○ ○
3.7	**E** Find the gradient and equation of a straight line perpendicular to a given line.	○ ○ ○

3.1 Cartesian coordinates

YOU NEED TO:

- Use and interpret Cartesian coordinates in two dimensions.

RECAP

Coordinates are pairs of numbers that determine a location on a grid, relative to a fixed point called the 'origin'. The origin has the coordinates (0, 0) and is the point where the two axes cross: the horizontal axis is the x-axis, and the vertical axis is the y-axis.

In a two-dimensional coordinate pair, the first number tells you the horizontal distance from the origin (the distance along the x-axis), and the second number tells you the vertical distance (the distance along the y-axis).

KEY SKILLS

You must be able to use and interpret two-dimensional coordinates.

WORKED EXAMPLE

What are the coordinates of the four points A, B, C, and D?

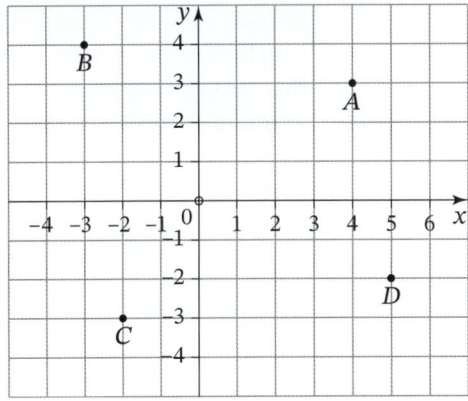

[4 marks]

To get to the point A from the origin, you go across by 4 and up by 3. This means that A has the coordinates (4, 3). The coordinates of B, C and D are (−3, 4), (−2, −3) and (5, −2) respectively.

EXAM TIP

Remembering that the letter x looks like 'a cross' (across) is a common way of remembering which axis is which.

EXAM TIP

Notice from the way the axes are labelled that positive coordinates are to the right and upwards, and negative coordinates are to the left and downwards.

Coordinates are always written inside round brackets with a comma separating the two numbers.

QUESTIONS

1. What are the coordinates of the points E, F, G, H, I and J?

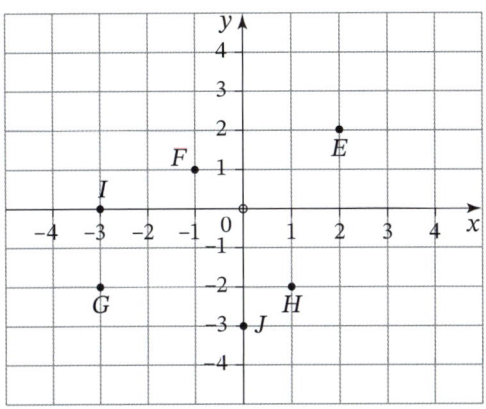

99

3 Coordinate geometry

3.2 Drawing straight-line graphs

YOU NEED TO:

- Draw straight-line graphs for linear equations.

WORKED EXAMPLE

Draw the line $y = 5 + 2x$. **[3 marks]**

This line has a y-intercept of 5 and a gradient of 2.

Set up a table of three values:

x	0	2	3
y	5	9	11

Plot the points and draw a straight line through them.

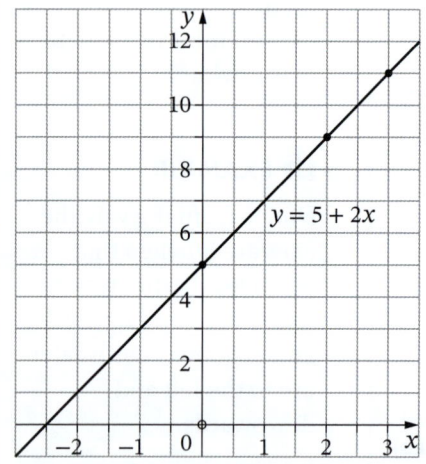

KEY SKILLS

You must be able to draw a straight-line graph from its equation.

EXAM TIP

Plot three points, not two, in case you make a mistake with one of them.

EXAM TIP

Sometimes you will need to rearrange the equation into the form $y = mx + c$ so that you can read off the gradient and y-intercept. This will also help you when substituting in values of x and y.

? QUESTIONS

1. Draw these lines.

    a) $y = 8 - \dfrac{2}{3}x$

    b) $3x - 5y = 15$

3.3 Gradients

YOU NEED TO:
- Find the gradient of a straight line with the aid of a grid.
- **E** Calculate the gradient of a straight line from the coordinates of two points on it.

RECAP

To find the gradient of a straight line, choose two points (x_1, y_1) and (x_2, y_2) on the line and divide the difference between the two y-coordinates by the difference between the two x-coordinates: gradient $= \dfrac{y_2 - y_1}{x_2 - x_1}$

KEY SKILLS

You must be able to find the gradient of a straight line, using two points on the line.

WORKED EXAMPLE

Find the gradient of the line that passes through the points (2, 1) and (4, 5).

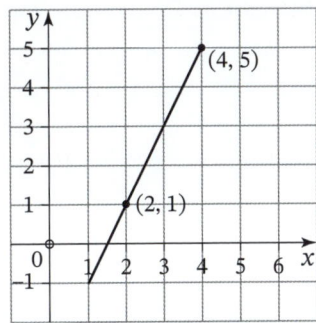

[2 marks]

Here $(x_1, y_1) = (2, 1)$ and $(x_2, y_2) = (4, 5)$

The gradient $= \dfrac{y_2 - y_1}{x_2 - x_1} = \dfrac{5 - 1}{4 - 2} = \dfrac{4}{2} = 2$

APPLY

Note that in the above worked example (x_1, y_1) was chosen as (2, 1) and (x_2, y_2) as (4, 5), but this is an arbitrary choice. It could have been the other way around. Recalculate the gradient with the coordinates the other way round and comment on your answer.

EXAM TIP

Usually, you can choose which point to subtract from the other based on reducing the amount of negative numbers involved in the calculation.

3 Coordinate geometry

? QUESTIONS

1. Find the gradient of each straight line in the diagram.

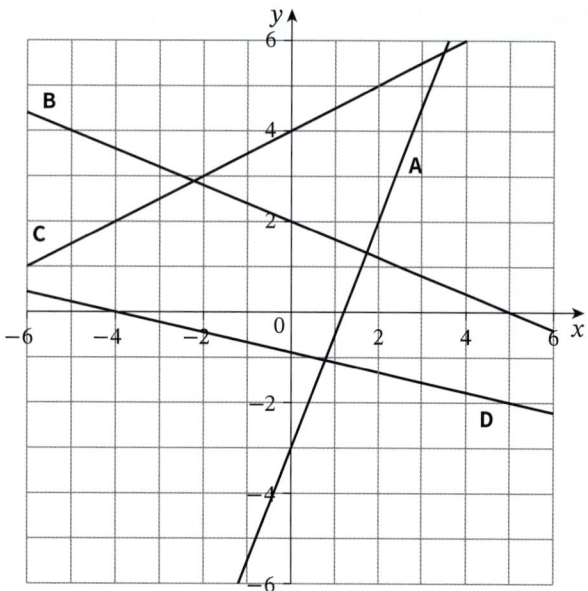

EXAM TIP

In Extended you will need to be able to find a gradient from two coordinates, whereas in Core you will always be shown a grid.

Extended

2. For each pair of points, find the gradient of the straight line that joins them.
    a) (1, 2) and (4, 6)
    b) (5, 7) and (3, 1)
    c) (−1, 4) and (−3, 9)
    d) (−2, 7) and (−5, −1)
    e) (−5, −6) and (−8, −11)
    f) (−10, −12) and (14, 0)

3. Find the value of a, given that the line joining (5, a) and (3, 10) has a gradient of −3.

4. Find the value of b, given that the line joining (1, 6) and (b, 2) has a gradient of $-\frac{1}{2}$.

5. Find the value of c, given that the line joining (−7, −2) and (−6, c) has a gradient of −3.

6. Find the value of d, given that the line joining (d, 6) and (3, d) has a gradient of 2.

7. Find the gradient of the line passing through the points (4e, 12e) and (7e, 9e).

8. The line through (5, p) and (3, 2) and the line through (2, −4) and (6, p) have the same gradient.
    a) Find the value of p.
    b) Find the gradient of each of these lines.

To **Raise your grade** now try questions 1 and 2 on page 110

3.4 Midpoints and lengths of lines

YOU NEED TO:

- **E** Calculate the length of a line segment.
- **E** Find the coordinates of the midpoint of a line segment.

Extended

◄◄ RECAP

To find the length of a line segment joining two points (x_1, y_1) and (x_2, y_2), use the formula:

distance $= \sqrt{(x_2 - x_1)^2 + (y_2 - y_1)^2}$

◄◄ RECAP

To find the midpoint of a line segment joining two points (x_1, y_1) and (x_2, y_2), use the formula:

midpoint $= \left(\dfrac{x_1 + x_2}{2}, \dfrac{y_1 + y_2}{2} \right)$

🖩 WORKED EXAMPLE

Point A is $(2, 1)$ and point B is $(8, 5)$.

a) Find the midpoint M of the line segment AB. **[2 marks]**

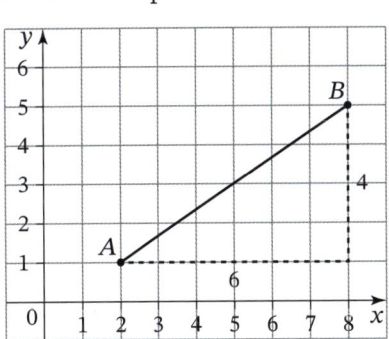

b) Find the length of AB. **[2 marks]**

a) The midpoint $M = \left(\dfrac{2+8}{2}, \dfrac{1+5}{2} \right) = (5, 3)$

b) The length of $AB = \sqrt{(8-2)^2 + (5-1)^2}$
$= \sqrt{36 + 16} = \sqrt{52} = 7.21$ (to 3 s.f.)

🗝 KEY SKILLS

You need to be able to calculate the length of a line segment joining two points, using the coordinates of the two points.

🗝 KEY SKILLS

You need to be able to find the midpoint of a line segment joining two points, using the coordinates of the two points.

👍 EXAM TIP

Part **a)** is the same as taking the mean of the two x-coordinates and the mean of the two y-coordinates.

Part **b)** is just a rearrangement of Pythagoras' theorem, where the line segment joining the two points is the hypotenuse of a triangle.

The triangle in the diagram shows this. See section 6.2 for revision of Pythagoras' theorem.

3 Coordinate geometry

> **QUESTIONS**
>
> 1. Point A has coordinates (4, 6), and point B is at (−2, 3).
>   a) Find the length of the line segment AB.
>   b) The midpoint of AB is point M. Find the coordinates of M.
> 2. A has coordinates (5, 4). The midpoint of the line segment AB is (1, 1).
>   a) Find the coordinates of B.
>   b) Find the length of the line segment AB.
>
> To **Raise your grade** now try question 6 on page 110

3.5 $y = mx + c$

YOU NEED TO:
- Interpret and obtain the equation of a straight-line graph in the form $y = mx + c$.

⏪ RECAP

Equations of straight lines

The equation of any non-vertical straight line can be written in the form

$y = mx + c$

where m is the gradient and c is the y-intercept.

The gradient, m, is a measure of how steep the line is.

$m = \dfrac{\text{change in } y}{\text{change in } x}$

m is positive if the line slopes upwards from left to right.

m is negative if the line slopes downwards from left to right.

The y-intercept of a straight line is the y-coordinate of the point where the line crosses the y-axis.

WORKED EXAMPLE

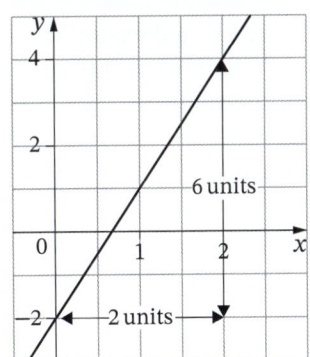

a) Write down the y-intercept of the line. **[1 mark]**
b) Find the gradient of the line. **[2 marks]**
c) Hence, write down the equation of the line. **[1 mark]**

a) The y-intercept is −2.

b) gradient $= \dfrac{\text{change in } y}{\text{change in } x} = \dfrac{4 - (-2)}{2 - 0} = \dfrac{6}{2} = 3$

c) $y = 3x - 2$

🔑 KEY SKILLS

You must be able to interpret the equation of a straight-line graph in the form $y = mx + c$.

👍 EXAM TIP

Straight lines with positive gradients go uphill as they move from left to right, so they look like this:

Straight lines with **N**egative gradients go downhill as they move from left to right, so they look like this:

Negative

3 Coordinate geometry

WORKED EXAMPLE

A line has the equation $y = 3x + 1$.

a) What is the gradient of the line? [1 mark]
b) Where does this line cross the y-axis? [1 mark]

$y = 3x + 1$ is of the form $y = mx + c$, with $m = 3$ and $c = 1$.

a) m is the gradient of the line, so the gradient of $y = 3x + 1$ is 3.
b) c is the y-intercept, so the line crosses the y-axis at $(0, 1)$.

👁 WATCH OUT!

Vertical lines have equations of the form $x = c$ and horizontal lines have equations of the form $y = c$, where c is a constant. Students often get these the wrong way around.

WORKED EXAMPLE

What is the equation of the straight line that passes through the points $(1, 3)$ and $(4, 9)$?

Give your answer in the form $y = mx + c$. [3 marks]

The gradient $m = \dfrac{9 - 3}{4 - 1} = \dfrac{6}{3} = 2$, so the equation is $y = 2x + c$.

To work out the y-intercept, substitute one of the points into this equation.

When $x = 1$, $y = 3$
$$y = 2x + c$$
$$3 = 2 \times 1 + c$$
$$3 = 2 + c$$
$$c = 1$$

So the equation is $y = 2x + 1$.

🔑 KEY SKILLS

You must be able to obtain the equation of a straight-line graph in the form $y = mx + c$ using two points on the line.

👍 EXAM TIP

It doesn't matter which of the two points you substitute in to find c.

✏️ APPLY

In the worked example on the left, try working out the value of c by substituting $x = 4$ and $y = 9$ into $y = 2x + c$ instead of $x = 1$ and $y = 3$ to check that you get the same answer.

❓ QUESTIONS

1. What is the equation of the straight line that passes through the points $(-1, -2)$ and $(1, 4)$?

   Give your answer in the form $y = mx + c$.

2. Find an equation of the line that:
   a) passes through $(0, 6)$ and has a gradient of 2
   b) passes through $(2, 3)$ and has a gradient of -1
   c) passes through $(-2, -5)$ and has a gradient of $\dfrac{1}{2}$.

3.6 Parallel lines

YOU NEED TO:
- Find the gradient and equation of a straight line parallel to a given line.

◀◀ RECAP
Two lines that are parallel have the same gradient. So $y = 3x + 1$ and $y = 3x - 5$ are parallel lines since they both have gradient 3.

🔑 KEY SKILLS
You must be able to find the equation of a line parallel to another line, given the equation of the other line and a point that the line passes through.

WORKED EXAMPLE
Find the line parallel to $y = 3x - 2$ which crosses the y-axis at the point (0, 5). **[3 marks]**

A line parallel to $y = 3x - 2$ has gradient 3.

A line passing through (0, 5) has y-intercept 5.

So the equation of the line is $y = 3x + 5$.

👍 EXAM TIP
Remember that when an equation is in the form $y = mx + c$, the gradient is m.

WORKED EXAMPLE
Find the line parallel to $y = 2x - 1$ which passes through the point (3, 7). **[3 marks]**

A line parallel to $y = 2x - 1$ has gradient 2 and therefore is of the form $y = 2x + c$.

If it passes through the point (3, 7), it must be true that $7 = 2 \times 3 + c$.

Therefore $c = 1$.

So the equation of the line is $y = 2x + 1$.

👍 EXAM TIP
If the point you are given is the y-intercept, this makes the question easier.

❓ QUESTIONS

1. Find the equation of the line parallel to $y = 5x + 10$ that passes through the point (0, 2).

2. Find the equation of the line parallel to $y = \frac{1}{2}x - 6$ that passes through the point (−2, 2).

3. Find the equation of the line that passes through (4, 1) and is parallel to the line $y = 6x - 2$.

4. Find the equation of the line parallel to $3x - 4y = 24$ that passes through the point (−4, 1).

   Give your answer in the form $ay - bx = c$, where a, b and c are all whole numbers.

   To **Raise your grade** now try question 4 on page 110

107

3 Coordinate geometry

3.7 Perpendicular lines

YOU NEED TO:

- **E** Find the gradient and equation of a straight line perpendicular to a given line.

Extended

⏪ RECAP

If two lines are perpendicular, the product of their gradients is −1:
$m_1 \times m_2 = -1$

WORKED EXAMPLE

Find the gradient of the lines perpendicular to these lines.

a) $y = 4x - 3$
b) $y = -0.5x + 1$
c) $2x + 3y = 12$ [4 marks]

a) Looking at $y = 4x - 3$, which is already in the form $y = mx + c$, the gradient is 4.
Therefore let $m_1 = 4$
$m_1 \times m_2 = -1$
so $m_2 = -1 \div m_1 = -1 \div 4 = \dfrac{1}{4}$

b) Looking at $y = -0.5x + 1$, which is already in the form $y = mx + c$, the gradient is −0.5.
Therefore let $m_1 = -0.5$
$m_1 \times m_2 = -1$
so $m_2 = -1 \div m_1 = -1 \div (-0.5) = 2$

c) First, rearrange the equation $2x + 3y = 12$ into the form $y = mx + c$.
$3y = -2x + 12$
$y = -\dfrac{2}{3}x + 4$, so $m_1 = -\dfrac{2}{3}$
The negative reciprocal of $-\dfrac{2}{3}$ is $\dfrac{3}{2}$, so $m_2 = \dfrac{3}{2}$.

🔑 KEY SKILLS

You must be able to find the gradient of a line that is perpendicular to a given line.

👍 EXAM TIP

It is often useful to rearrange an equation into the form $y = mx + c$ to find the gradient or y-intercept of the line.

👍 EXAM TIP

A quick way to find the perpendicular gradient is to write the original gradient as a fraction, then just turn the fraction upside down and change the sign. This is called finding the 'negative reciprocal'.

3.7 Perpendicular lines

WORKED EXAMPLE

Find the equation of the straight line through (3, 1) that is perpendicular to the line $y = 0.5x + 1$. **[3 marks]**

Gradient of the perpendicular line: $-1 \div 0.5 = -2$.

The equation of a straight line is $y = mx + c$ and $m = -2$, so $y = -2x + c$.

The line passes through (3, 1), so

$1 = -2 \times 3 + c$

$c = 7$

Hence the equation is $y = -2x + 7$.

KEY SKILLS

You must be able to find the equation of a line perpendicular to another line, given the equation of the other line and a point your line passes through.

QUESTIONS

1. Find the gradient of the lines perpendicular to the following lines.
   a) $y = 3x - 1$
   b) $y = -2x + 3$
   c) $y = \frac{1}{4}x - 5$
   d) $2x - 3y = 7$

2. Find the equation of the line that:
   a) passes through (2, 3) and is perpendicular to $y = x - 4$
   b) passes through (−3, 5) and is perpendicular to $y = -2x + 1$
   c) passes through (6, −1) and is perpendicular to $2y = 3x + 5$
   d) passes through (−1, −2) and is perpendicular to the line through points (2, 3) and (4, 6).

3. Find the equation of the straight line through (4, 2) that is perpendicular to the line $y = -2x + 3$.

4. a) In the diagram below, find the equation of the line CD.
   b) The line AB has the equation $2y - x = -2$.

   Show that the lines AB and CD are perpendicular.

5. Find the gradient of the line perpendicular to the line $2y - 5x = 11$.

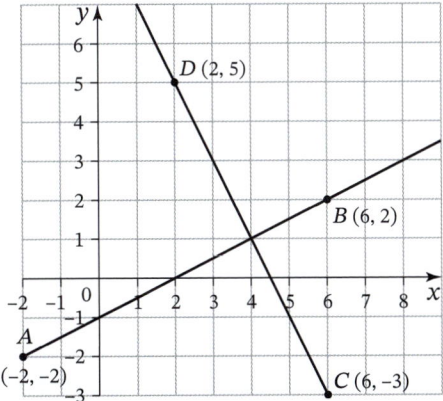

6. Two points A and B have coordinates (5, 7) and (−1, 3), respectively.
   a) Find the midpoint of the line segment AB.
   b) Find the gradient of AB.
   c) Find the equation of the perpendicular bisector of the line segment AB.

To **Raise your grade** now try questions 3 and 5 on page 110

3 Coordinate geometry

↑ Raise your grade

1. Find the value of j, given that the line joining points $(j, 2)$ and $(5, 7)$ has a gradient of -1. **[3 marks]**

2. Find the value of k, given that the line joining points $(3, k)$ and $(k, 5)$ has a gradient of 2. **[3 marks]**

Extended

3. Find the equation of the line that:
   a) passes through $(6, -1)$ and is perpendicular to $2y = 3x + 5$ **[3 marks]**
   b) passes through $(-1, -2)$ and is perpendicular to the line through $(2, 3)$ and $(4, 6)$. **[4 marks]**

4. a) Find the gradient of the line joining the points $(m, 3n)$ and $(2, -6)$. **[1 mark]**
   b) Find the value of m if the line is parallel to the y-axis. **[2 marks]**
   c) Find the value of n if the line is parallel to the x-axis. **[2 marks]**

5. A straight line passes through two points with coordinates $A(4, 5)$ and $B(8, 11)$. Find the equation of the straight line that passes through the midpoint of AB and is perpendicular to the line through A and B. **[7 marks]**

6. Point A has coordinates $(4, 6)$, and point B is at $(-2, 3)$.
   a) Find the length of the line segment AB. **[2 marks]**
   b) The midpoint of AB is point M. Find the coordinates of M. **[2 marks]**
   c) Find the coordinates of point P which divides AB in the ratio $2:1$. **[2 marks]**

4 Geometry

Your revision checklist

Tick these circles to build a record of your revision.

Core/ **E** Extended syllabus

☹	😐	🙂

4.1 Use and interpret the geometric terms: point, vertex, line, plane, parallel, perpendicular, bearing, right-angle, acute, obtuse and reflex angles, interior and exterior angles, similar, congruent and scale factor.

Use and interpret the vocabulary of triangles, special quadrilaterals, polygons, nets and solids.

E Use and interpret the geometric term perpendicular bisector.

Use and interpret the vocabulary of a circle.

4.2 Measure and draw lines and angles.

Construct a triangle given the three sides, using a ruler and pair of compasses only.

Read and make nets.

4.3 Read and make scale drawings.

Use and interpret three-figure bearings.

4.4 Calculate lengths of similar shapes.

E Use relationships between lengths and areas of similar shapes, and lengths, surface areas and volumes of similar solids.

E Solve similarity problems and give simple explanations of similarity.

4.5 Recognise line symmetry and order of rotational symmetry in two dimensions.

E Recognise symmetry properties of prisms (including cylinders) and pyramids (including cones).

4.6 Calculate unknown angles and give simple explanations using the following geometric properties:
- sum of angles at a point = 360°
- sum of angles at a point on a straight line = 180°
- vertically opposite angles are equal
- angle sum of a triangle = 180° and angle sum of a quadrilateral = 360°

Calculate unknown angles and give geometric explanations for angles formed within parallel lines:
- corresponding angles are equal
- alternate angles are equal
- co-interior (supplementary) angles sum to 180°

Know and use angle properties of regular polygons.

E Know and use angle properties of irregular polygons.

4.7 Calculate unknown angles and give explanations using the following geometrical properties of circles:
- angle in a semicircle = 90°
- angle between tangent and radius = 90°

E Calculate unknown angles and give explanations using the following geometrical properties of circles:
- angle at the centre of a circle is twice the angle at the circumference
- angles in the same segment are equal
- opposite angles in a cyclic quadrilateral sum to 180°
- alternate segment theorem

4.8 **E** Use the following symmetry properties of circles:
- equal chords are equidistant from the centre
- perpendicular bisector of a chord passes through the centre
- tangents from an external point are of equal length

4 Geometry

4.1 Geometric vocabulary

> **YOU NEED TO:**
> - Use and interpret the geometric terms: point, vertex, line, plane, parallel, perpendicular, bearing, right-angle, acute, obtuse and reflex angles, interior and exterior angles, similar, congruent, scale factor.
> - Use and interpret the vocabulary of triangles, special quadrilaterals, polygons, nets and solids.
> - **E** Use and interpret the geometric term perpendicular bisector.
> - Use and interpret the vocabulary of a circle.

◀◀ RECAP

An **acute angle** is less than 90°.

An **obtuse angle** is between 90° and 180°.

A **reflex angle** is larger than 180° but less than 360°.

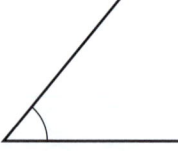

A **right-angle** is 90°.

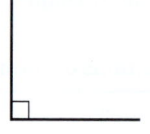

Extended

◀◀ RECAP

Perpendicular lines intersect at right angles.

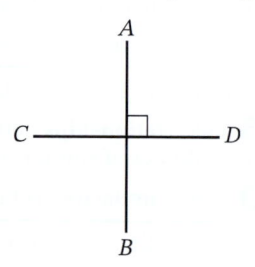

◀◀ RECAP

A **polygon** is a closed 2-dimensional shape with straight edges.

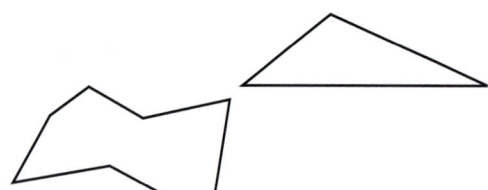

A **regular polygon** has all the sides equal and all the angles equal.

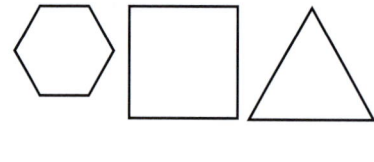

🗝 KEY SKILLS

At extended level, you are expected to know the definition of a perpendicular bisector. A perpendicular bisector is a line that divides a line segment into two equal parts and is at right angles to it.

◀◀ RECAP

Parallel lines never meet. Arrows are used to indicate parallel lines.

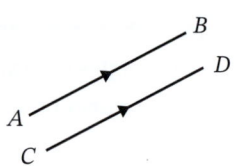

4.1 Geometric vocabulary

◀◀ RECAP

An **isosceles triangle** has two equal sides and two equal angles.

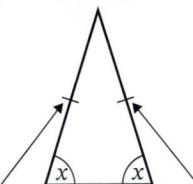

An **equilateral triangle** has three equal sides and three equal angles.

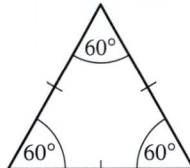

A **right-angled triangle** has one right-angle.

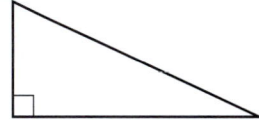

The marks indicate that the sides are equal.

◀◀ RECAP

A **quadrilateral** is a four-sided polygon.

Square

4 equal sides, 4 right angles

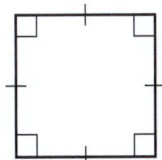

Rectangle

4 sides, 4 right angles, opposite sides equal

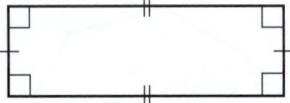

Parallelogram

4 sides, 2 sets of parallel sides, opposite sides equal, opposite angles equal

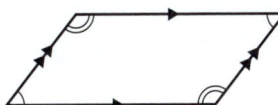

Rhombus

4 equal sides, opposite sides parallel, opposite angles equal

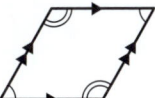

Trapezium

4 sides, one set of opposite sides parallel

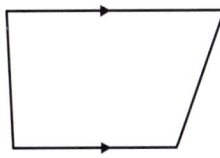

Kite

4 sides, 2 pairs of adjacent sides equal, 1 pair of opposite angles equal

◀◀ RECAP

Two shapes are **congruent** if the corresponding sides and the corresponding angles are equal.

✏ APPLY

Answer **yes** or **no** to these questions. Explain your answer.

- Is a square a rectangle?
- Is a parallelogram a rhombus?
- Is a rectangle a parallelogram?
- Is a rhombus a kite?

🔑 KEY SKILLS

You need to be able to recognise and use the geometrical properties of quadrilaterals and other polygons.

4 Geometry

QUESTIONS

1. Name these polygons.

   a)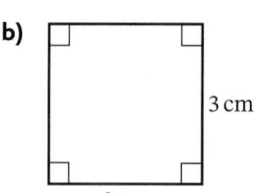

   b)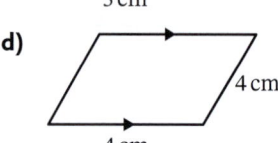

   c)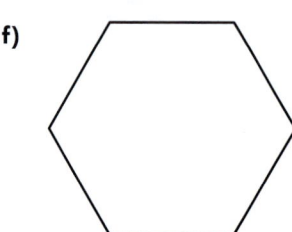

   d)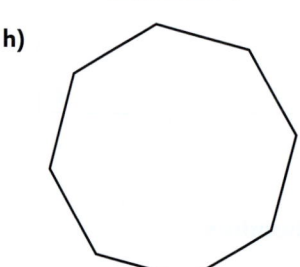

   e)

   f)

   g)

   h)

   i)

2. Identify pairs of congruent shapes.

   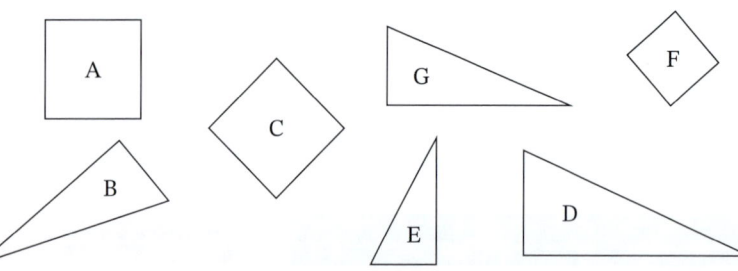

4.2 Constructions and nets

YOU NEED TO:
- Measure and draw lines and angles.
- Construct a triangle, given the lengths of all three sides, using a ruler and pair of compasses only.
- Read and make nets.

WORKED EXAMPLE

Construct a triangle with side lengths 5 cm, 6 cm and 8 cm. **[3 marks]**

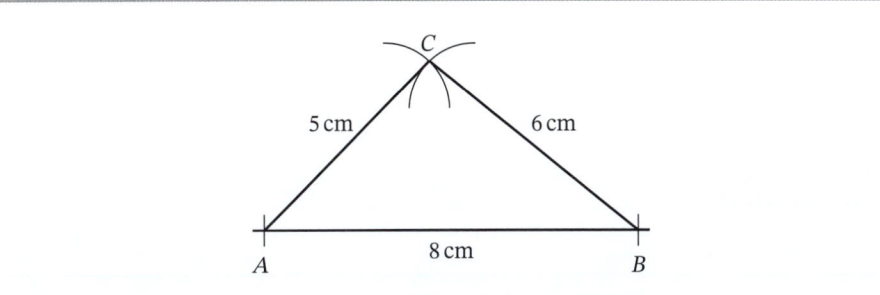

◀◀ RECAP

To construct a triangle given the three side lengths, you use a ruler and pair of compasses.

👍 EXAM TIP

Draw the longest side first using a ruler.

Then draw little arcs of radius 5 cm and 6 cm above your line.

Where the arcs cross will be the third corner of the triangle, so you can draw the other sides.

◀◀ RECAP

If you cut along the edges of a 3-dimensional shape and lay it out flat, you have a **net**.

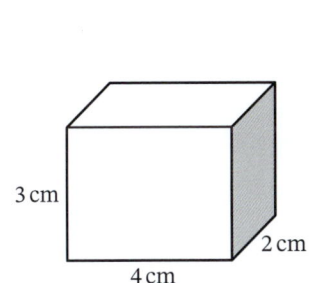 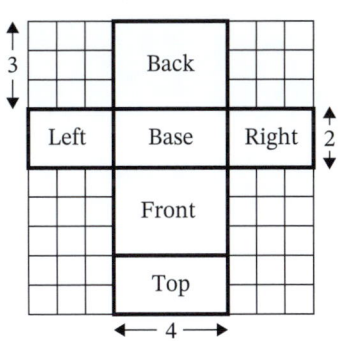

❓ QUESTIONS

1. Construct an equilateral triangle *ABC* where *AB* is:
   a) 5 cm
   b) 3.7 cm

2. Make an accurate full-sized copy of this diagram.

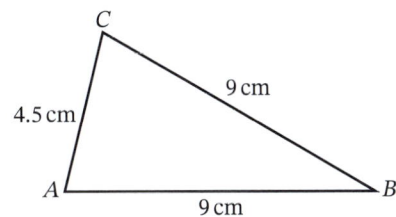

3. Make an accurate full-sized copy of this diagram.

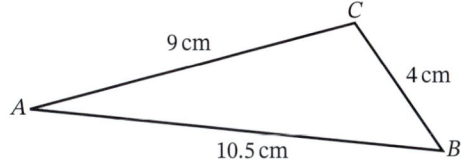

4. Explain why it is not possible to construct a triangle *ABC* where *AB* = 10 cm, *BC* = 5 cm and *CA* = 4 cm.

5. Make an accurate full-sized copy of this diagram.

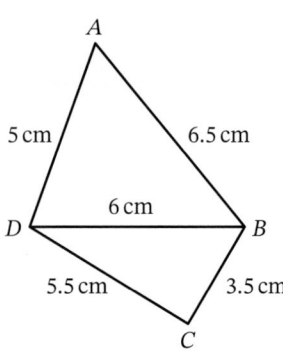

7. Sketch the net of a cuboid with side lengths 3 cm, 5 cm and 7 cm.

8. Using a scale of 1 cm to 2 m, draw a net of this isosceles triangular prism.

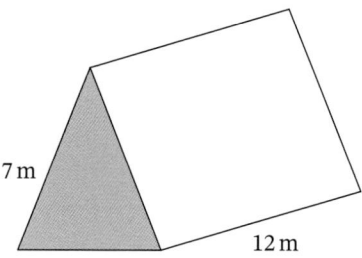

6. Which two of the nets below can be used to make a cube?

a)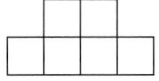

b)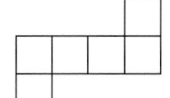

c)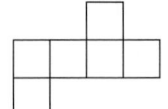

d)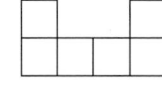

4.3 Bearings

> **YOU NEED TO:**
> - Read and make scale drawings.
> - Use and interpret three-figure bearings.

WORKED EXAMPLE

A triangular field ABC has sides $AB = 120\,\text{m}$ and $AC = 80\,\text{m}$.

Angle $BAC = 40°$.

By making an accurate scale drawing of the field, find the length of side BC. **[3 marks]**

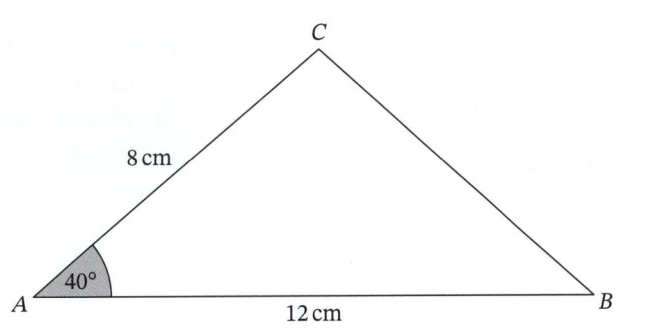

$BC = 7.8\,\text{cm}$

So the length of side BC of the triangular field is $78\,\text{m}$.

KEY SKILLS

You need to be able to make an accurate scale drawing from information given in either a diagram or text.

EXAM TIP

Use a sensible scale, for example, $1\,\text{cm} = 10\,\text{m}$.

Measure the angle BAC using a protractor.

Measure the side length of your scale drawing and use your scale to write down the actual length of side BC.

RECAP

Bearings tell you a direction as an angle measured clockwise from the north line.

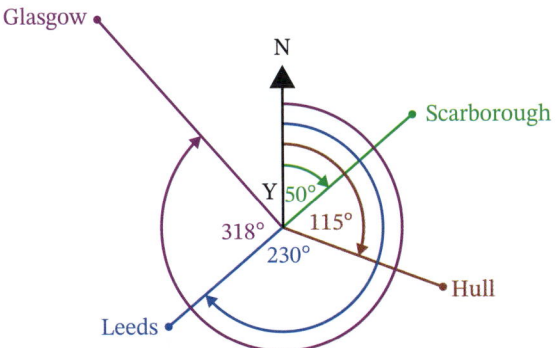

Scarborough is on a bearing of 050° from York (Y).

Hull is on a bearing of 115° from York.

Leeds is on a bearing of 230° from York.

Glasgow is on a bearing of 318° from York.

4 Geometry

WORKED EXAMPLE

The bearing of a ship from a lighthouse is 100°.

Find the bearing of the lighthouse from the ship. **[2 marks]**

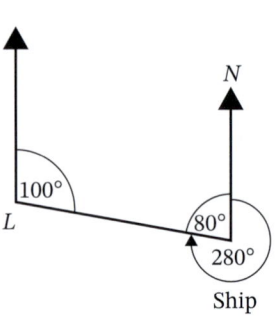

Acute angle $LSN = 180° - 100° = 80°$

Bearing $= 360° - 80° = 280°$

KEY SKILLS

If you are given the bearing of B from A, you need to be able to find the bearing of A from B.

This is known as the **back bearing**.

EXAM TIP

Draw a little diagram and use angle rules.

EXAM TIP

A quick way of working out the back bearing is to add 180° if the initial bearing is less than 180° or subtract 180° if it is greater than 180°.

WORKED EXAMPLE

Clinton walks towards Abi's house on a bearing of 240°.

Write down the bearing on which Abi should walk if she goes to meet Clinton. **[1 mark]**

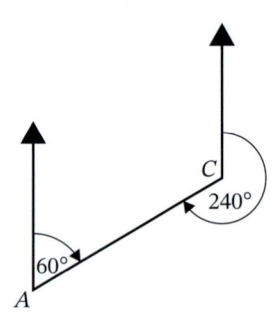

$240° - 180° = 060°$

? QUESTIONS

1. A garden has four sides, as shown in the diagram.

   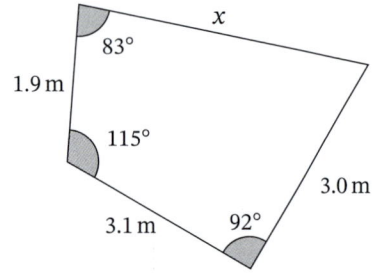

   Use an accurate scale drawing to find the length marked x.

2. The bearing of Cambridge from Oxford is 065°. Work out the bearing of Oxford from Cambridge.

3. A plane flies from an airport on a bearing of 320°. Work out the bearing on which it needs to fly to get back to the airport.

4. Use a scale drawing to work out the answers to each of these questions.

   a) A ship sails 300 km on a bearing of 080° and then a further 250 km on a bearing of 110°. Find how far the ship is from where it started, and the bearing on which it should sail to get directly back to port.

b) Two ports, A and B, are 60 km apart and B is due south of A.

A yacht is on a bearing of 095° from A and on a bearing of 036° from B.

A speedboat is on a bearing of 240° from A and on a bearing of 315° from B.

How far apart are the yacht and speedboat?

c) Two cats, Fluffy and Harbury, are playing in a field.

They start from the same place.

Fluffy runs at a constant speed of $4\,\text{m s}^{-1}$ on a bearing of 054°.

Harbury runs at a constant speed of $5\,\text{km h}^{-1}$ on a bearing of 285°.

How far apart are the cats after 10 seconds?

5. The diagram shows each of three landmarks in a village.

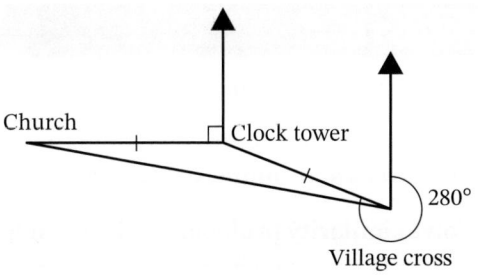

The bearing of the clock tower from the village cross is 280°.

The church is due West of the clock tower.

The distance from the village cross to the clock tower is the same as the distance from the clock tower to the church.

Calculate the bearing of the village cross from the church.

4 Geometry

4.4 Similar shapes

YOU NEED TO:
- Calculate lengths of similar figures.
- **E** Use relationships between lengths and areas of similar shapes, and lengths, surface areas and volumes of similar solids.
- **E** Solve similarity problems and give simple explanations of similarity.

◀◀ RECAP

Two shapes are **similar** if the ratio of every pair of corresponding sides is the same.

You can think of one of the shapes as being an **enlargement** of the other.

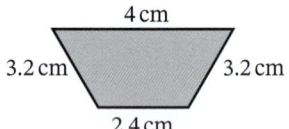
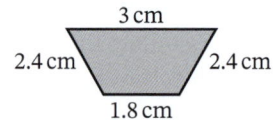

In the two shapes above, all of the corresponding side length ratios are equal to 0.75 so the shapes are similar.

WORKED EXAMPLE

These two shapes are similar.

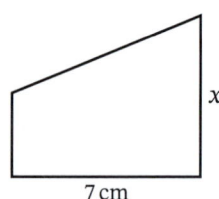
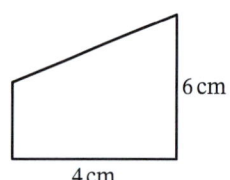

Find x. **[2 marks]**

🔑 KEY SKILLS

You need to be able to find a missing side length when you are given two similar shapes.

$\dfrac{7}{4} = 1.75$

Hence $x = 1.75 \times 6 = 10.5$ cm

WORKED EXAMPLE

Triangles ABC and ADE are similar.

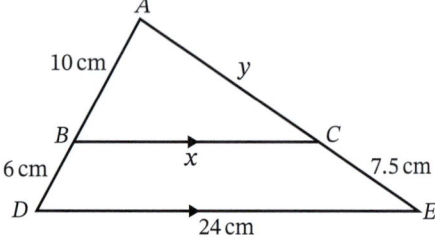

Find x and y. **[4 marks]**

Triangles *ABC* and *ADE*

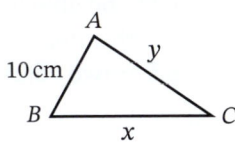
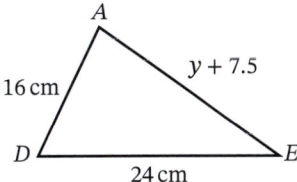

$\dfrac{16}{10} = 1.6$

Hence $x = 24 \div 1.6 = 15\,\text{cm}$

$\dfrac{y + 7.5}{y} = 1.6 \Rightarrow y + 7.5 = 1.6y$

$7.5 = 0.6y \Rightarrow y = 12.5\,\text{cm}$

> **EXAM TIP**
>
> Draw the two similar triangles separately and mark on all of the lengths before calculating the ratio of corresponding sides.

Extended

WORKED EXAMPLE

These two triangles are similar.

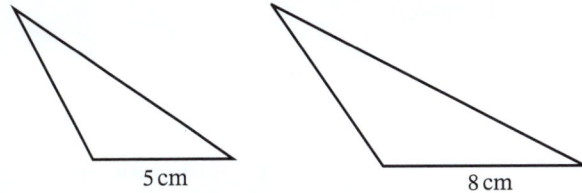

The area of the smaller triangle is $24\,\text{cm}^2$.

Find the area of the larger triangle. **[3 marks]**

Ratio of side lengths = $\dfrac{8}{5}$

Area = $24 \times \left(\dfrac{8}{5}\right)^2 = 61.44\,\text{cm}^2$

WORKED EXAMPLE

Two similar cylinders have radius 3 cm and 5 cm, respectively.

The volume of the smaller cylinder is $54\,\text{cm}^3$.

Find the volume of the larger cylinder. **[3 marks]**

Ratio of side lengths = $\dfrac{5}{3}$

Volume = $54 \times \left(\dfrac{5}{3}\right)^3 = 250\,\text{cm}^3$

> **RECAP**
>
> If two shapes are similar, the ratio of their areas is equal to the square of the ratio of their sides.

> **EXAM TIP**
>
> Multiply the area of the smaller triangle by the square of the ratio of side lengths.

> **RECAP**
>
> If two solids are similar, the ratio of their surface areas is equal to the square of the ratio of their sides, and the ratio of their volumes is equal to the cube of the ratio of their sides.

> **EXAM TIP**
>
> Multiply the volume of the smaller cylinder by the cube of the ratio of side lengths.

4 Geometry

WORKED EXAMPLE

Two bottles are similar.

Bottle A has volume 1200 cm³ and height 10 cm.

Bottle B has volume 1900 cm³.

Find the height of bottle B. **[3 marks]**

Ratio of volumes = $\frac{1900}{1200} = \frac{19}{12}$

Ratio of side lengths = $\sqrt[3]{\frac{19}{12}}$

Hence height of bottle B = $10 \times \sqrt[3]{\frac{19}{12}} = 11.7$ cm (3 s.f.)

🔑 KEY SKILLS

You need to be able to find a side length when given two areas or two volumes.

You also need to be able to find an area when given two volumes, and find a volume when given two areas.

👍 EXAM TIP

If the ratio of volumes is the cube of the ratio of lengths, the ratio of lengths is the cube root of the ratio of volumes.

WORKED EXAMPLE

Two containers are similar.

Container D has a volume of 8000 cm³ and a surface area of 5000 cm².

Container E has a volume of 4000 cm³.

Find the surface area of container E. **[3 marks]**

Ratio of volumes = $\frac{4000}{8000} = 0.5$

Ratio of side lengths = $\sqrt[3]{0.5}$

Ratio of surface areas = $\left(\sqrt[3]{0.5}\right)^2$

Hence surface area of E = $5000 \times \left(\sqrt[3]{0.5}\right)^2 = 3150$ cm² (3 s.f.)

👍 EXAM TIP

Use the ratio of volumes to work out the ratio of side lengths and then convert this to the ratio of areas.

❓ QUESTIONS

1. Ben and Sarah want to measure the height of a building.

   Ben is 1.8 m tall and Sarah suggests that he stands next to the building and compares the shadows.

   She measures his shadow to be 2.4 m long and the shadow of the building to be 16 m long.

   How tall is the building?

2. A photocopier is set to reduce the lengths of copies to $\frac{2}{3}$ of the original size.

   An original document measures 12 cm by 15 cm.

   Work out the dimensions of the copy.

3. A photography shop produces enlargements of photos.

   A 15 cm × 10 cm photo was enlarged so that its longest side was 24 cm.

   Work out the length of the shorter side.

4. A map is reduced to $\frac{3}{5}$ of its original size.

   A field on the original map measured 25 mm by 35 mm.

   Work out the dimensions on the image.

5. A map that measures 24 cm by 30 cm is reduced to $\frac{2}{3}$ of its original size.

   Work out the dimensions of the reduced map. →

6. In the triangle in the diagram, $BD = 8\,cm$, $AB = 10\,cm$, $AD = 6\,cm$, $AC = x$ and $CD = y$.

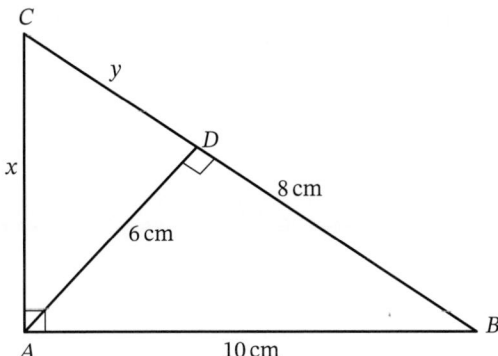

a) Draw the two triangles ABC and DBA in the same orientation and mark on all their angles.
b) Hence explain why triangles ABC and DBA are similar.
c) Write down an equation involving x.
d) Solve the equation to find x.
e) Find the value of y.

7. A rectangle P is enlarged to a rectangle Q.
The dimensions of P are 5 m by 12 m.
The shortest side of Q is 6 m.
a) Work out the scale factor of enlargement.
b) Work out the length of the longer side of Q.

8. A right-angled triangle P is enlarged to triangle Q.
The hypotenuse of P is 12 cm and the hypotenuse of Q is 15 cm.
a) Work out the scale factor of enlargement.
The shortest side of P is 8 cm.
b) Work out the length of the shortest side of Q.

9. A photo 8 cm high and 10 cm wide has a border that is 2 cm high along the bottom and the top of the photo and that is w cm wide on each side.
The original photo is similar to the photo with its border.
Find the value of w.

10. In the diagram, $\angle DCB = \angle CAB = \theta$, $DB = 8\,cm$, $DC = 12\,cm$ and $CB = 10\,cm$.

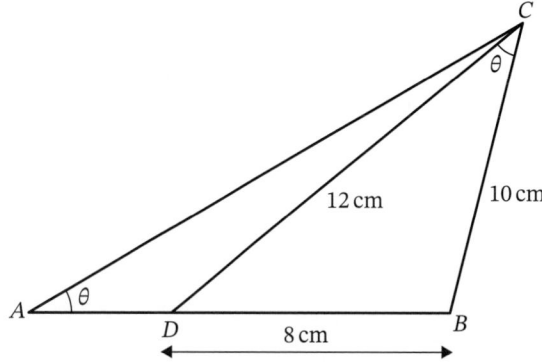

a) Which triangle is similar to triangle ABC?
b) Find the length AB.
c) Find the length AC.

11. A cone of radius 6 cm and height 15 cm has a cone of height 9 cm removed from its top.
Work out the radius of the removed cone.

12. The distance between Delhi and Calcutta is 1310 km.
On a map they are 26.2 cm apart.
Find the scale of the map in the form $1:n$.

Extended

13. The scale of a map is $1:20\,000\,000$.
On the map the area of a state is $5\,cm^2$.
Calculate the actual area of the state, giving your answer in km^2.

14. In the diagram, $AB = 5\,cm$, $BC = 4\,cm$ and the area of the triangle ABE is $23\,cm^2$.

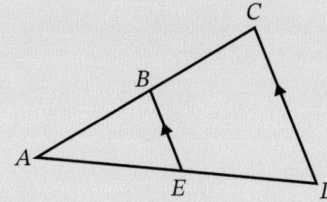

BE is parallel to CD.
Work out the area of the triangle ACD.
Give your answer to 2 significant figures.

15. Two large water tanks are similar.
One holds $5\,m^3$ and the other holds $12\,m^3$.
The height of the smaller tank is 1.2 m.
Work out the height of the larger tank.
Give your answer to 3 significant figures.

16. Two pictures, A and B, are similar.
The area of A is $54\,cm^2$ and the area of B is $216\,cm^2$.
The length of B is 18 cm.
Work out the length of A.

17. Three layers of a wedding cake are similar.

The middle layer has a surface area of 3600 cm² and a mass of 5 kg.

The bottom layer has a surface area of 8000 cm².

a) Work out the mass of the bottom layer.

The mass of the top layer is 3 kg.

b) Work out the surface area of the top layer.

18. Two cuboids C and D are similar.

Cuboid C has volume 6 m³ and cuboid D has volume 11 m³.

The surface area of C is 22 m².

Work out the surface area of D.

Give your answer to 3 significant figures.

To **Raise your grade** now try question 4 on page 139

4.5 Symmetry

> **YOU NEED TO:**
> - Recognise line symmetry and order of rotational symmetry in two dimensions.
> - **E** Recognise symmetry properties of prisms (including cylinders) and pyramids (including cones).

◄◄ RECAP

An equilateral triangle has three lines of symmetry.

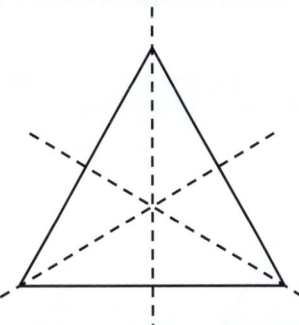

◄◄ RECAP

A square has four lines of symmetry.

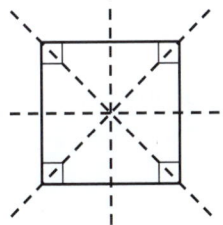

A rectangle has two lines of symmetry.

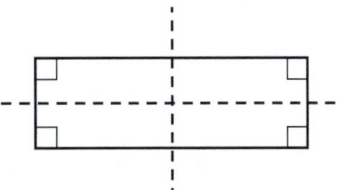

A parallelogram has no lines of symmetry.

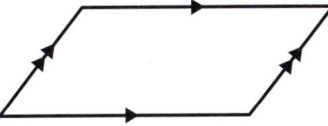

A rhombus has two lines of symmetry.

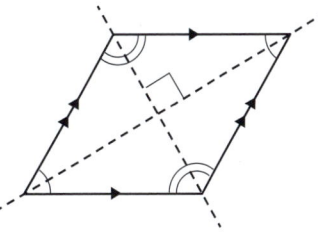

A trapezium has no lines of symmetry.

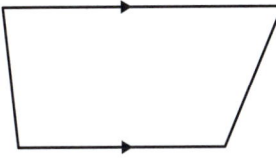

A kite has one line of symmetry.

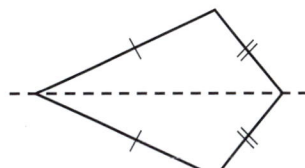

4 Geometry

> **◀◀ RECAP**
>
> A regular *n*-sided polygon has *n* lines of symmetry.
>
>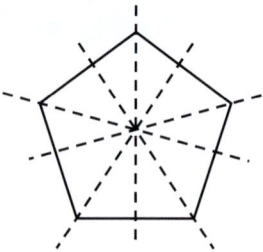
>
> For example, a regular pentagon has five lines of symmetry.

> **◀◀ RECAP**
>
> An equilateral triangle has rotational symmetry of order 3.
>
>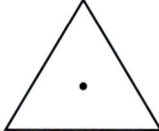
>
> A square has rotational symmetry of order 4.   A rectangle has rotational symmetry of order 2.
>
> 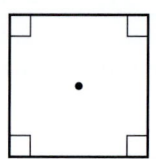   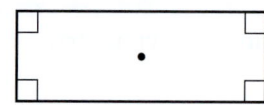
>
> A parallelogram has rotational symmetry of order 2.   A rhombus has rotational symmetry of order 2.
>
> 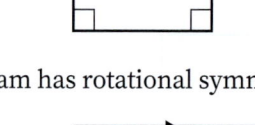   
>
> A trapezium has rotational symmetry of order 1.   A kite has rotational symmetry of order 1.
>
> 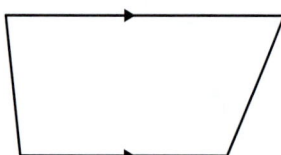   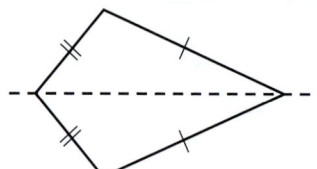
>
> A regular *n*-sided polygon has rotational symmetry of order *n*.

4.5 Symmetry

WORKED EXAMPLE

For these shapes:

a) draw all the lines of symmetry
b) write down the order of rotational symmetry. **[4 marks]**

i)    ii)

a) Lines of symmetry:

i) 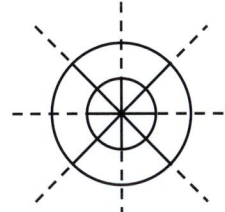   ii)

4 lines of symmetry   8 lines of symmetry

b) Order of rotational symmetry:
   i) 4
   ii) 8

🔑 KEY SKILLS

You need to be able to draw lines of symmetry, state the number of lines of symmetry and find the order of rotational symmetry.

👍 EXAM TIP

Imagine placing a mirror on the line of symmetry. The reflected part of the shape should appear identical.

For shape i), rotating by 90°, 180°, 270° and 360° gives the same shape.

Extended

⏪ RECAP

A cylinder and cone both have an **axis** of symmetry.

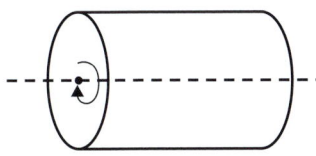   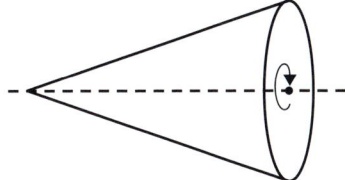

Prisms with a regular n-sided polygon cross-section have order of rotational symmetry n.

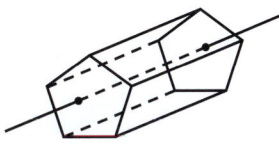

For example, a pentagonal prism has order of rotational symmetry 5.

4 Geometry

⏮ RECAP

A cuboid has, in general, three **planes of symmetry**.

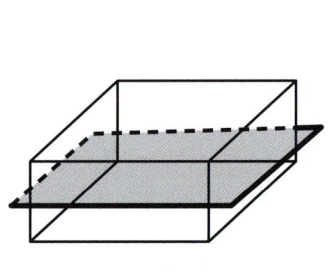 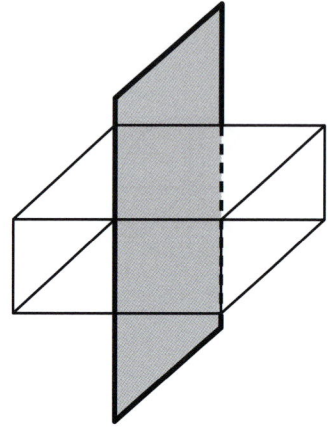 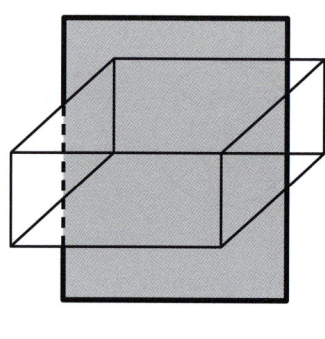

Prisms with a regular *n*-sided polygon cross-section have *n* + 1 planes of symmetry.

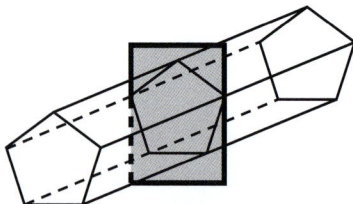 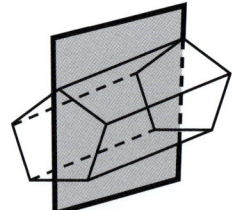

For example, a pentagonal prism has 6 planes of symmetry.

✏ APPLY

If a cuboid has one pair of opposite faces that are square, then it will have five planes of symmetry.

Sketch a cuboid with this feature and draw the planes of symmetry.

How many planes of symmetry does a cube have?

WORKED EXAMPLE

a) On a copy of the diagram, sketch one of the three planes of symmetry of this cuboid. **[1 mark]**

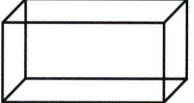

b) Write down the order of rotational symmetry of an equilateral triangular prism. **[1 mark]**

a) This is one of the planes of symmetry

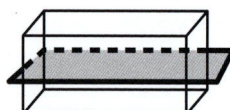

b) The order of rotational symmetry is 3.

👍 EXAM TIP

You could have sketched either of the other two planes of symmetry.

👍 EXAM TIP

A prism with a regular *n*-sided polygon cross-section has order of rotational symmetry *n*.

128

4.5 Symmetry

? QUESTIONS

1. Copy the following shapes and:
   a) draw on all the lines of symmetry
   b) state the order of rotational symmetry.

2. For this shape, write down:
   a) the order of rotational symmetry
   b) the number of lines of symmetry.

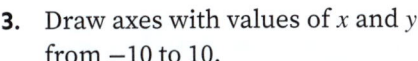

3. Draw axes with values of x and y from -10 to 10.

   Draw these shapes and fill in a copy of the table.

	Shape	Vertices	Missing vertices	Lines of symmetry
a)	Rectangle	(1, 1), (5, 1), (5, 3)	(__, __)	$x = 3$, $y = $ __
b)	Rectangle	(6, 1), (6, 9)	(__, __) (__, __)	$x = 8$, $y = 5$
c)	Isosceles triangle	(3, 5), (5, 5), (4, 9)		$x = $ __
d)	Isosceles triangle	(0, 5), (3, 7)	(__, __)	$y = 7$ only
e)	Rhombus	(−3, 0), (−1, 1), (0, 3)	(__, __)	$y = -x$, $y = $ __
f)	Trapezium	(3, −4), (2, −7), (7, −7)	(__, __)	$x = 4.5$ only
g)	Parallelogram	(−6, −6), (−5, −3), (−2, −3)	(__, __)	None

4. Shade one square in each diagram so that there is:

   a) one line of symmetry

   b) rotational symmetry of order 2.

Extended

5. The cuboid shown below has no square faces.

   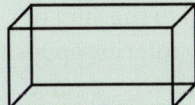

   How many planes of symmetry does it have?

6. The diagram shows a regular hexagonal prism.

   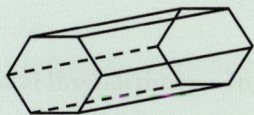

   a) Write down the number of planes of symmetry.
   b) Write down the order of rotational symmetry.

7. How many planes of symmetry does a right square-based pyramid have?

4 Geometry

4.6 Angles

YOU NEED TO:

- Calculate unknown angles and give simple explanations using the following geometric properties:
  - sum of angles at a point = 360°
  - sum of angles at a point on a straight line = 180°
  - vertically opposite angles are equal
  - angle sum of a triangle = 180° and angle sum of a quadrilateral = 360°
- Calculate unknown angles and give geometric explanations for angles formed within parallel lines:
  - corresponding angles are equal
  - alternate angles are equal
  - co-interior (supplementary) angles sum to 180°
- Know and use angle properties of regular polygons.
- **E** Know and use angle properties of irregular polygons.

◀◀ RECAP

Angles at a point add up to 360°.

$a + b + c + d = 360°$

Angles on a straight line add up to 180°.

$a + c = b + d = 180°$

Vertically opposite angles are equal.

$a = b$ and $c = d$

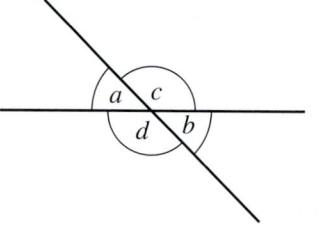

◀◀ RECAP

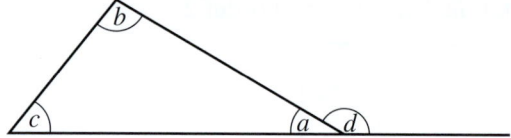

The angle sum in a triangle is 180°.

$a + b + c = 180°$

The exterior angle of a triangle is equal to the sum of the two interior opposite angles.

$d = b + c$

The angle sum of a quadrilateral is 360°.

✎ APPLY

Construct a proof that the exterior angle of a triangle is equal to the sum of the two interior opposite angles.

4.6 Angles

◀ RECAP

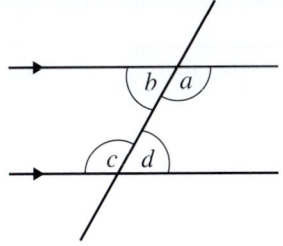

Alternate angles are equal.

$a = c$ and $b = d$

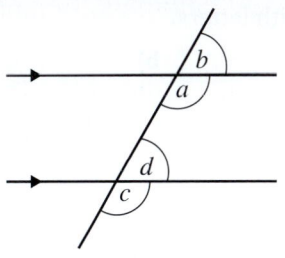

Corresponding angles are equal.

$a = c$ and $b = d$

Co-interior (supplementary) angles add up to 180°.

$a + d = 180°$

WORKED EXAMPLE

Find the angles marked with letters in the diagram below.

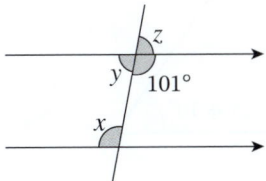

Give a reason for each answer. **[4 marks]**

$x = 101°$ (Alternate angles are equal)

$y = 180° - 101° = 79°$ (Angles on a straight line add up to 180°)

$z = 79°$ (Opposite angles are equal)

🔑 KEY SKILLS

You need to be able to use angle rules to find unknown angles.

👍 EXAM TIP

State your reasons clearly using the correct terminology.

◀ RECAP

The **exterior angle** of a regular n-sided polygon is $\dfrac{360°}{n}$

The **interior angle** of polygon = 180° − exterior angle

Extended

◀ RECAP

The sum of the interior angles in an n-sided polygon is $(n - 2) \times 180°$.

For example, the sum of the interior angles of a 7-sided polygon is $5 \times 180 = 900°$.

4 Geometry

? QUESTIONS

1. Find the angles marked with letters.

   a)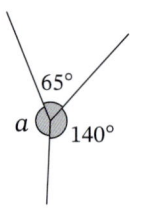

   b)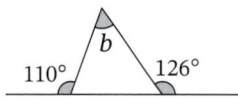

   c)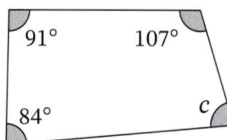

   d)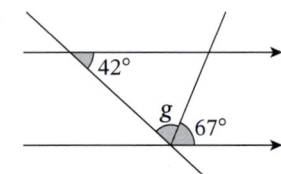

   e)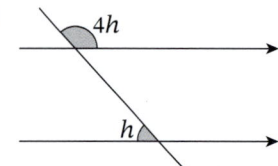

2. Find the value of x in these quadrilaterals.

   a)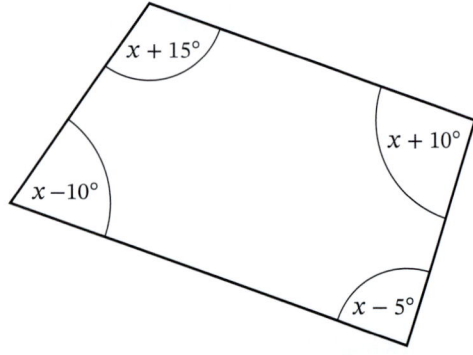

   b)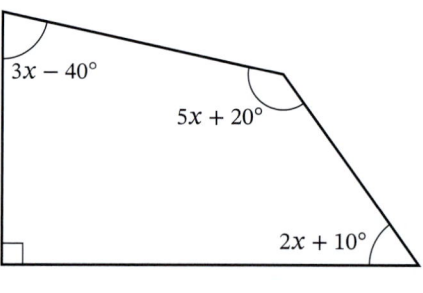

3. Calculate the largest angle in a quadrilateral in which one angle is 5 times each of the other angles.

4. In an isosceles trapezium, the smaller angle is equal to 53°.
   Find the larger angle.

5. A triangle has angles in the ratio 2 : 3 : 4.
   Find the size of the largest angle.

6. In a regular polygon, each exterior angle is twice the size of each interior angle.
   Find the number of sides.

7. Name a polygon in which the sum of the interior angles is:

   a) 180°   b) 540°   c) 360°   d) 720°

8. The sum of the interior angles of a polygon is 1440°.
   Find the number of sides.

9. Find the angles marked with letters.

   a)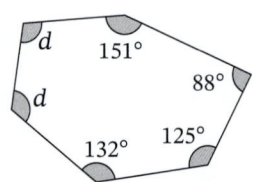

   b)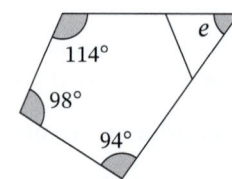

To **Raise your grade** now try questions 1, 2 and 3 on page 139

4.7, 4.8 Circle theorems

> **YOU NEED TO:**
>
> - Calculate unknown angles and give explanations using the following geometrical properties of circles:
>   - angle in a semicircle = 90°
>   - angle between tangent and radius = 90°
>
> - **E** Calculate unknown angles and give explanations using the following geometrical properties of circles:
>   - angle at the centre of a circle is twice the angle at the circumference
>   - angles in the same segment are equal
>   - opposite angles in a cyclic quadrilateral sum to 180°
>   - alternate segment theorem
>
> - **E** Use the following symmetry properties of circles:
>   - equal chords are equidistant from the centre
>   - the perpendicular bisector of a chord passes through the centre
>   - tangents from an external point are of equal length

◀◀ RECAP

PR is a diameter of the circle and point Q lies on the circumference.

The angle in a semicircle is 90°.

Angle $RQP = 90°$.

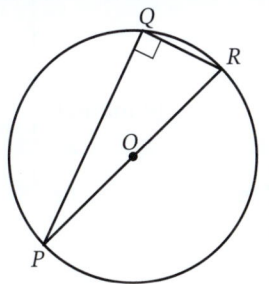

◀◀ RECAP

OR is a radius of the circle and RT is a tangent to the circle at R.

The angle between a radius and a tangent is 90°.

Angle $ORT = 90°$.

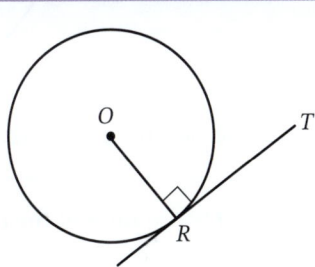

Extended

◀◀ RECAP

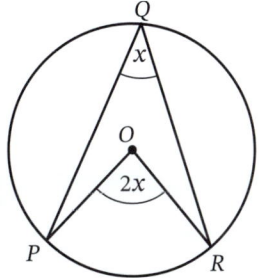

The angle at the centre of a circle is twice the angle at the circumference.

⊙ WATCH OUT!

This rule works even if the angles are not in the shape of an 'arrowhead'.

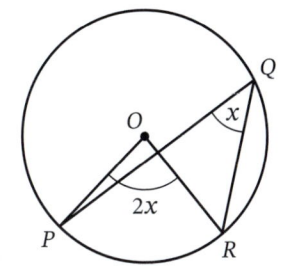

133

4 Geometry

⏪ RECAP

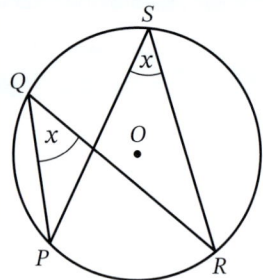

Angles in the same segment are equal.

WORKED EXAMPLE

Find the values of x and y in this diagram.

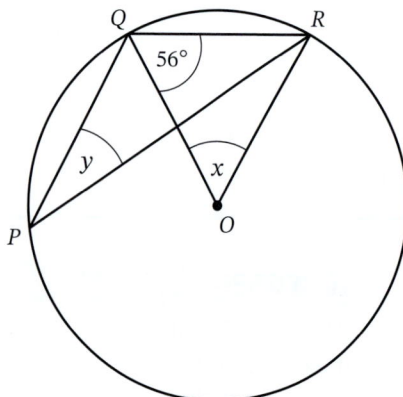

Give a reason for each stage of your working. **[4 marks]**

OQ and OR are radii of the circle so triangle OQR is isosceles.

Therefore, angle $QRO = 56°$

$x = 180° - 2 \times 56° = 68°$ (Angles in a triangle add up to 180°)

$y = \dfrac{68°}{2} = 34°$ (Angle at the centre is twice the angle at the circumference)

🔑 KEY SKILLS

You need to be able to use circle theorems to find missing angles.

👍 EXAM TIP

There will be marks in the exam for clear explanations and reasons.

Make sure you use the correct terminology too.

⏪ RECAP

A **cyclic quadrilateral** has all four vertices on the circumference of a circle.

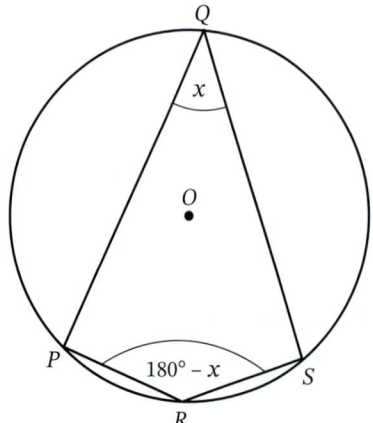

Opposite angles in a cyclic quadrilateral add up to 180°.

4.7, 4.8 Circle theorems

WORKED EXAMPLE

Find the values of x, y and z in this diagram.

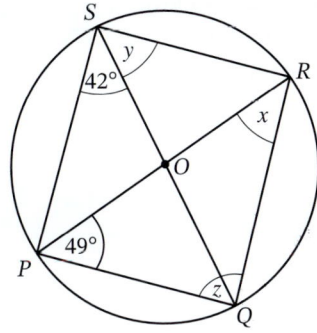

Give a reason for each stage of your working. **[4 marks]**

$x = 42°$ (angles in the same segment are equal)

$y = 49°$ (angles in the same segment are equal)

$z = 180° - (42° + 49°) = 89°$ (opposite angles in a cyclic quadrilateral add up to 180°)

👍 EXAM TIP
Make sure to show each stage of your working and give clear reasons.

⏪ RECAP

The angle between a tangent and a chord is equal to the angle in the alternate segment.

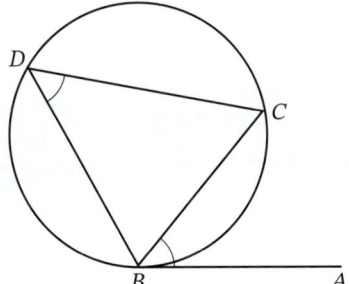

(This is known as the **alternate segment theorem**.)

WORKED EXAMPLE

Find the angle marked g in this diagram.

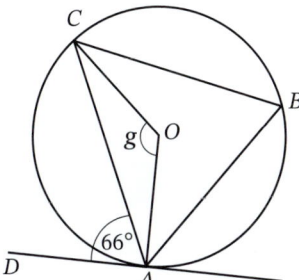

Give a reason for each stage of your working. **[3 marks]**

Angle $ABC = 66°$ (Alternate segment theorem)

$g = 66° \times 2 = 132°$ (Angle at the centre is twice the angle at the circumference)

👍 EXAM TIP
You can write 'Alternate segment theorem' as a reason without explaining it any further.

135

4 Geometry

⏪ RECAP

Equal chords

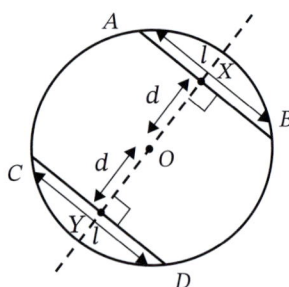

If two chords AB and CD have equal length, then they are the same perpendicular distance from the centre of the circle.

$OX = OY$

Bisector of chord

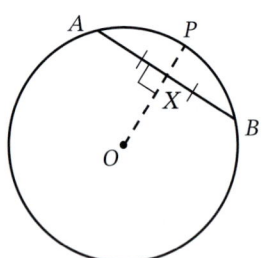

The perpendicular line from the centre of a circle to a chord bisects the chord.

$AX = XB$

⏪ RECAP

Tangents

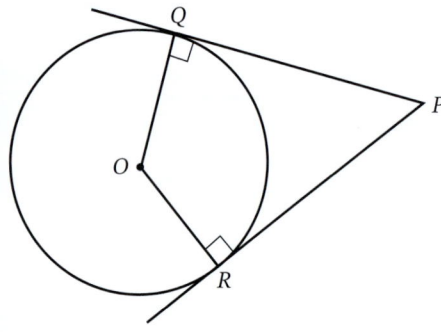

The tangents from a point P outside a circle to two points Q and R on the circle are equal in length.

$PQ = PR$

> 👍 **EXAM TIP**
>
> These facts are usually used in the context of questions on circle theorems or in questions on Pythagoras' theorem and trigonometry.

4.7, 4.8 Circle theorems

QUESTIONS

1. Find, giving reasons, the angles shown by letters.

a)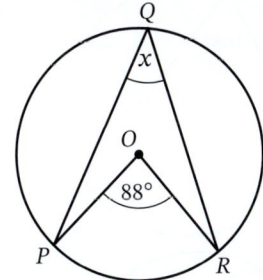

b)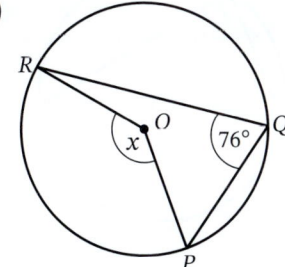

c)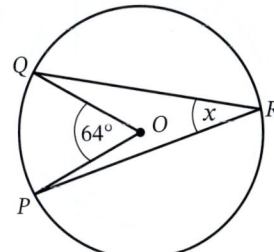

d)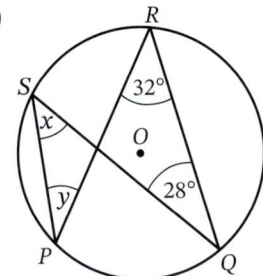

e)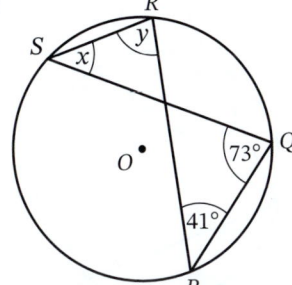

f)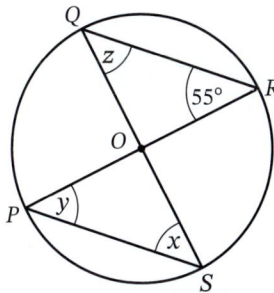

g)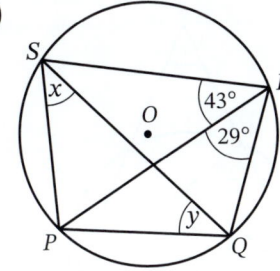

h)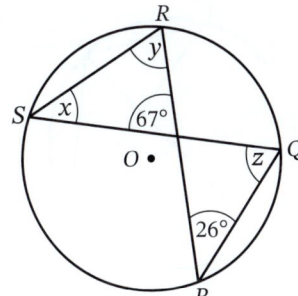

i)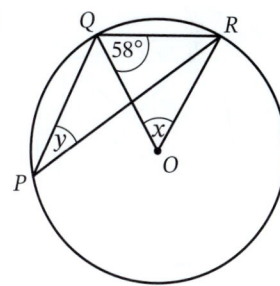

j)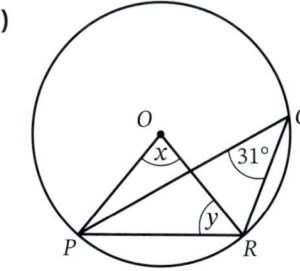

k)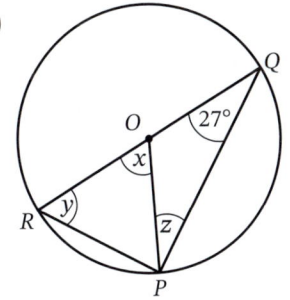

l)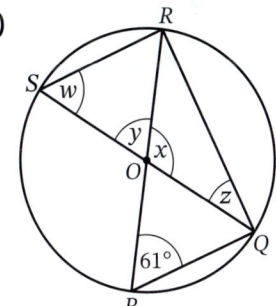

4 Geometry

2. Find, giving reasons, the angles shown by letters.

a)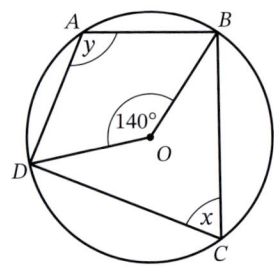

b)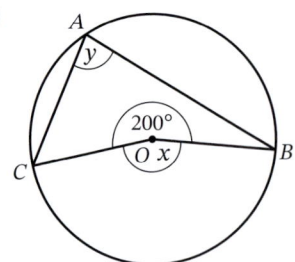

c)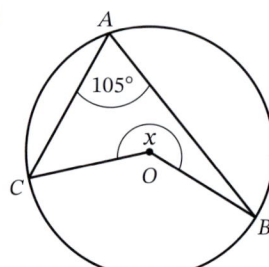

d)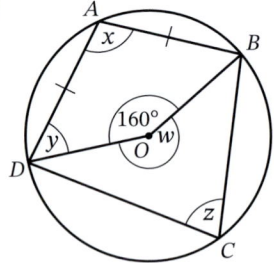

e)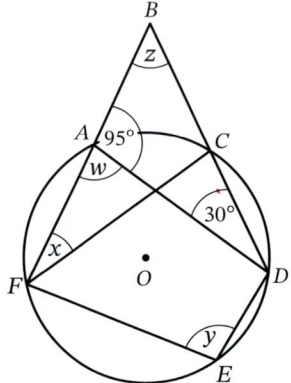

f)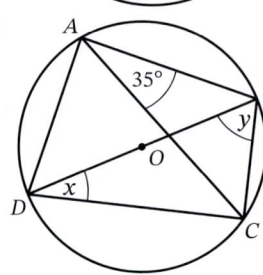

g)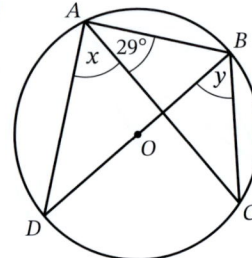

h)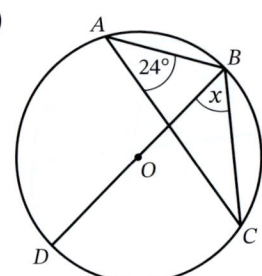

i)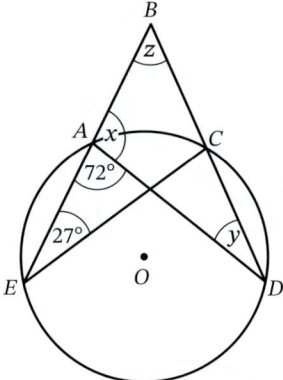

3. Find, giving reasons, the angles shown by letters.

a)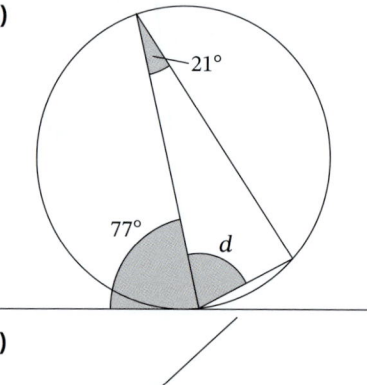

b)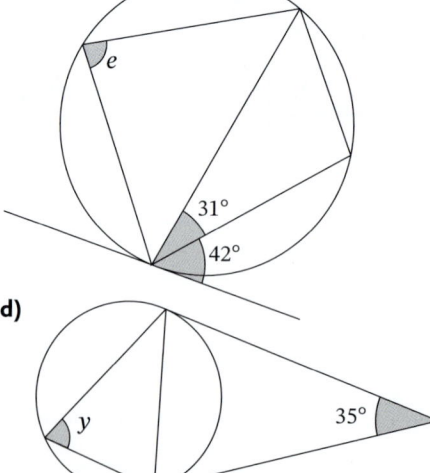

c)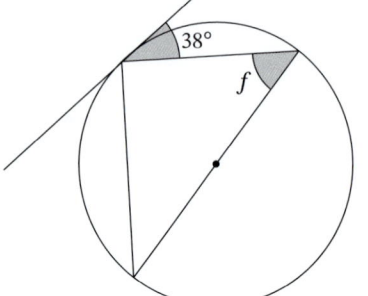

d)

To **Raise your grade** now try question 5 on page 139

↑ Raise your grade

1. A regular decagon rests on side AB.

   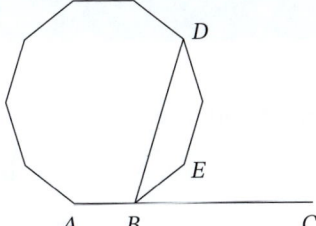

   ABC is a straight line.

   Calculate:

   a) angle EBC [2 marks]

   b) angle ABD. [3 marks]

2. A triangle has angles in the ratio 2 : 3 : 4.

   Find the size of the largest angle. [3 marks]

3. Find the size of angle x and the size of angle y.

   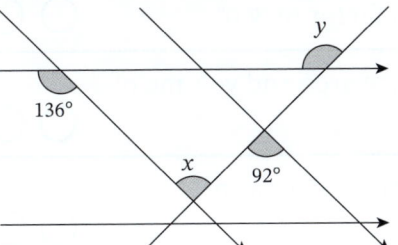

   [4 marks]

Extended

4. Two similar barrels have heights in the ratio 6 : 5.

   The surface area of the larger barrel is $2.4 \, m^2$.

   a) Find the surface area of the smaller barrel. [3 marks]

   The volume of the smaller barrel is $1.2 \, m^3$.

   b) Find the volume of the larger barrel. [3 marks]

5. A, B and C lie on the circumference of a circle with centre O.

   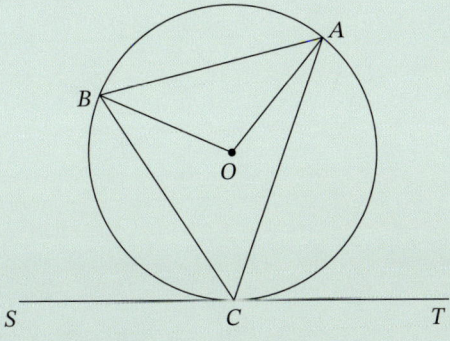

   ST is straight line, and it is tangent to the circle at C.

   Reflex angle $BOA = 250°$ and angle $ABC = 82°$

   Find:

   a) angle BCA [3 marks]

   b) angle BCS. [3 marks]

5 Mensuration

Your revision checklist

Tick these circles to build a record of your revision.

Core/ E **Extended** syllabus

5.1	Use current units of mass, length, area, volume and capacity in practical situations and express quantities in terms of larger or smaller units.
5.2	Carry out calculations involving the perimeter and area of a rectangle, triangle, parallelogram and trapezium, and compound shapes made from these.
5.3	Carry out calculations involving the circumference and area of a circle.
	Carry out calculations involving arc length and sector area as fractions of the circumference and area of a circle where the sector angle is a factor of 360°.
5.4	Carry out calculations and solve problems involving the surface area and volume of a cuboid, prism, cylinder, sphere, pyramid and cone.
5.5	Carry out calculations and solve problems involving perimeters and areas of compound shapes and parts of shapes.
	Carry out calculations and solve problems involving surface areas and volumes of compound shapes and parts of solids.

5.1 Units of measurement

YOU NEED TO:
- Use current units of mass, length, area, volume and capacity in practical situations and express quantities in terms of larger or smaller units.

RECAP

Units of length

$10\,mm = 1\,cm$

$100\,cm = 1\,m$

$1000\,m = 1\,km$

Units of mass

$1000\,g = 1\,kg$

$1000\,kg = 1\,tonne$

Units of capacity

$1\,cm^3 = 1\,ml$

$1000\,ml = 1\,litre$

Units of area

$100\,mm^2 = 1\,cm^2$

$10\,000\,cm^2 = 1\,m^2$

$1\,000\,000\,m^2 = 1\,km^2$

Units of volume

$1000\,mm^3 = 1\,cm^3$

$1\,000\,000\,cm^3 = 1\,m^3$

WATCH OUT!

Units of area and units of volume have different conversions to the standard units of length.

It is useful to **learn** these conversions and remember to use them in the exam.

WORKED EXAMPLE

Calculate the number of square centimetres in 5 square metres. **[2 marks]**

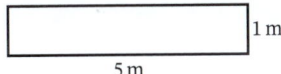

$1\,m \times 5\,m = 100\,cm \times 500\,cm$

$= 50\,000\,cm^2$

EXAM TIP

Imagine a rectangle with dimensions such that the area is 5 square metres.

Alternatively, use the conversion from m^2 to cm^2 so $5 \times 10\,000 = 50\,000$.

QUESTIONS

1. Complete these unit conversions.
   a) $1\,m^2 = \ldots\ldots\ldots\ldots\,cm^2$
   b) $1\,km^2 = \ldots\ldots\ldots\ldots\,m^2$
   c) $20\,000\,cm^2 = \ldots\ldots\ldots\ldots\,m^2$
   d) $50\,000\,m^2 = \ldots\ldots\ldots\ldots\,km^2$
   e) $14.8\,km^2 = \ldots\ldots\ldots\ldots\,m^2$
   f) $1\,380\,000\,cm^2 = \ldots\ldots\ldots\ldots\,m^2$
   g) $0.25\,m^2 = \ldots\ldots\ldots\ldots\,cm^2$
   h) $0.00436\,km^2 = \ldots\ldots\ldots\ldots\,m^2$
   i) $0.00000547\,km^2 = \ldots\ldots\ldots\ldots\,mm^2$

2. Complete these unit conversions.
   a) $1\,cm^3 = \ldots\ldots\ldots\ldots\,mm^3$
   b) $1\,m^3 = \ldots\ldots\ldots\ldots\,cm^3$
   c) $200\,000\,cm^3 = \ldots\ldots\ldots\ldots\,m^3$
   d) $6500\,mm^3 = \ldots\ldots\ldots\ldots\,cm^3$
   e) $1000\,cm^3 = \ldots\ldots\ldots\ldots\,litres$
   f) $1\,m^3 = \ldots\ldots\ldots\ldots\,litres$
   g) $6.7\,litres = \ldots\ldots\ldots\ldots\,cm^3$
   h) $1\,km^3 = \ldots\ldots\ldots\ldots\,m^3$

5 Mensuration

5.2 Perimeter and area

YOU NEED TO:
- Carry out calculations involving the perimeter and area of a rectangle, triangle, parallelogram and trapezium, and compound shapes made from these.

◀◀ RECAP

Rectangle

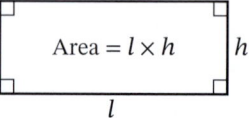

Perimeter $= 2l + 2h$

Triangle

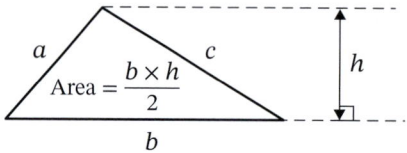

Perimeter $= a + b + c$

Parallelogram

The area of the parallelogram is the same as the area of the dashed rectangle.

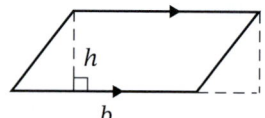

Area of parallelogram $= b \times h$

Trapezium

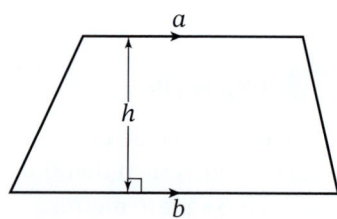

Area of trapezium is $\frac{1}{2}(a + b) \times h$

WORKED EXAMPLE

Work out the area and perimeter of each of these shapes.

a) [2 marks]

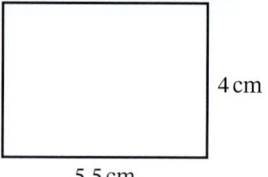

b) [2 marks]

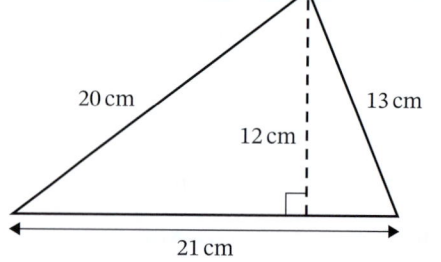

c) [2 marks]

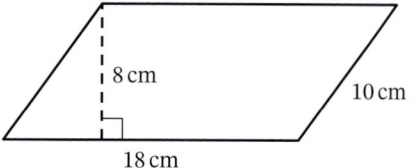

d) [3 marks]

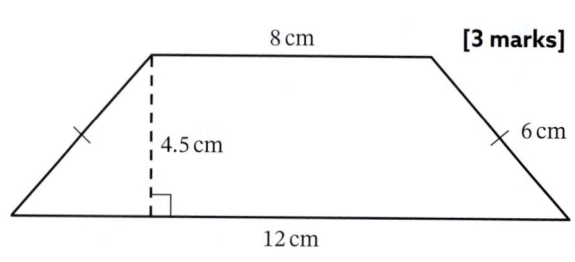

a) Area = 4 × 5.5 = 22 cm²
    Perimeter = 2 × 4 + 2 × 5.5 = 19 cm
c) Area = 18 × 8 = 144 cm²
    Perimeter = 2 × 18 + 2 × 10 = 56 cm

b) Area = $\frac{1}{2}$ × 21 × 12 = 126 cm²
    Perimeter = 21 + 13 + 20 = 54 cm
d) Area = $\frac{1}{2}$(8 + 12) × 4.5 = 45 cm²
    Perimeter = 12 + 8 + 2 × 6 = 32 cm

? QUESTIONS

1. Find the area and perimeter of these shapes.

   a) Rectangle

   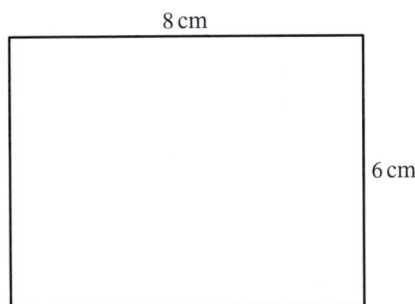

   b) Parallelogram

   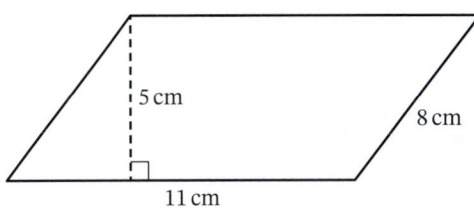

2. Find the area of these shapes.

   a) Triangle

   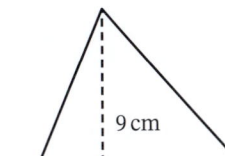

   b) Trapezium

   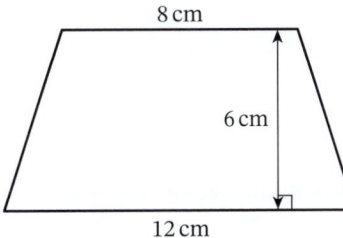

   c) Kite

   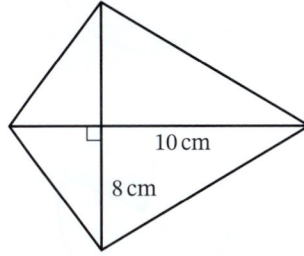

3. A rectangle has area 126 cm² and width 4.5 cm.
   Calculate:
   a) the length of the rectangle
   b) the perimeter of the rectangle.

4. A triangle has area 52 cm² and base 2.6 cm.
   Calculate the height of the triangle.

5. A trapezium of area 210 cm² has parallel sides of length 10 cm and 18 cm.
   How far apart are the parallel sides?

To **Raise your grade** now try question 7 on page 153

5 Mensuration

5.3 Circles

YOU NEED TO:
- Carry out calculations involving the circumference and area of a circle.
- Carry out calculations involving arc length and sector area as fractions of the circumference and area of a circle, where the sector angle is a factor of 360°.

⏮ RECAP

Circle

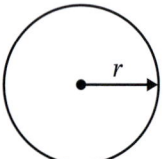

Area = πr^2

Circumference = $2\pi r$

Sector of circle

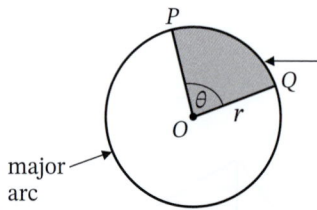

Length of minor arc = $\dfrac{\theta}{360} \times 2\pi r$

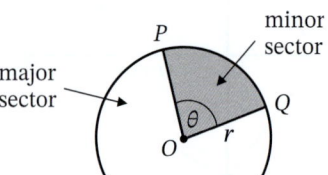

Area of minor sector = $\dfrac{\theta}{360} \times \pi r^2$

🔑 KEY SKILLS

You need to be able to use the formulas for the length of an arc and the area of a sector.

WORKED EXAMPLE

A large circular cheese has radius 12 cm.

A wedge is cut from the cheese in the shape of a sector with angle 40°.

Calculate:

a) the cross-sectional area of the wedge
b) the total perimeter of the wedge. [5 marks]

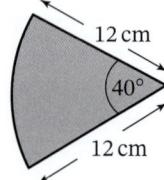

a) Area of sector = $\dfrac{40}{360} \times \pi \times 12^2 = 16\pi = 50.3 \text{ cm}^2$ (3 s.f.)

b) Length of arc = $\dfrac{40}{360} \times 2 \times \pi \times 12 = \dfrac{8}{3}\pi = 8.38 \text{ cm}$ (3 s.f.)

Perimeter = $\dfrac{8}{3}\pi + 2 \times 12 = \dfrac{8}{3}\pi + 24 = 32.4 \text{ cm}$ (3 s.f.)

👍 EXAM TIP

Don't forget to add on the two radii for the total perimeter.

Use the exact answer for the arc length to ensure that no rounding errors are carried forwards.

👁 WATCH OUT!

You may be asked for the answer as a rounded decimal or as a multiple of π.

If it doesn't specify, it is best to give both.

144

5.3 Circles

? QUESTIONS

1. Find, in terms of π, the area that is enclosed between two circles with the same centre, one with radius 7 cm and the other with radius 5 cm.

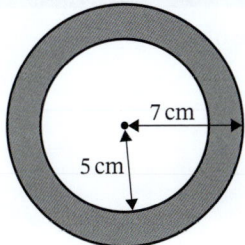

2. Find the radius of the circle which has the same area as the combined area of a circle of radius 12 cm and a circle of radius 5 cm.

3. A circle has an area of 100 cm².

   Work out the radius.

   Give your answer correct to 3 significant figures.

4. A circle has a circumference of 50 m.

   Work out the area.

   Give your answer correct to 3 significant figures.

5. Find the area and perimeter of these sectors, correct to 3 significant figures.

   a) Acute sector

   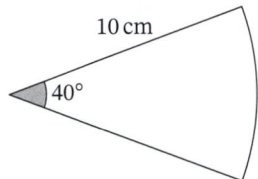

   b) Reflex sector

   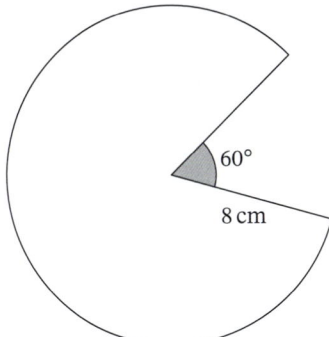

6. The perimeter of this sector is 14.33 cm.

   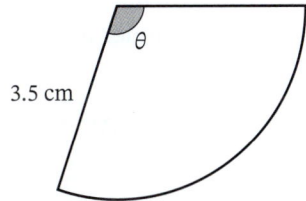

   Find the value of θ.

   To **Raise your grade** now try question 6 on page 153

5 Mensuration

5.4 3D shapes

YOU NEED TO:
- Carry out calculations and solve problems involving the surface area and volume of a cuboid, prism, cylinder, sphere, pyramid and cone.

◀◀ RECAP

Cuboid

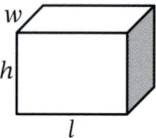

Volume = $w \times l \times h$

Surface area = $2hl + 2hw + 2wl = 2(hl + hw + wl)$

Prism

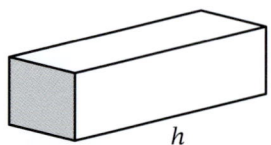 

Volume = cross-sectional area × height

WATCH OUT!
There is no formula as such for the surface area of a prism.

It depends on the shape and dimensions of the cross-section.

◀◀ RECAP

Cylinder

A cylinder is a prism with a circular cross-section.

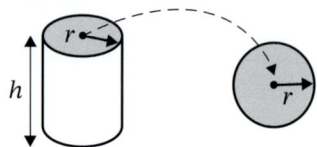

Cross-sectional area = πr^2

Volume = $\pi r^2 h$

The **curved surface area** of a cylinder = $2\pi rh$ (or πdh where d is the diameter)

The **total surface area** of a cylinder = $2\pi rh + 2\pi r^2$

WATCH OUT!
Always read the question carefully to ensure that you calculate the correct surface area.

WORKED EXAMPLE

🖩 A cylindrical can of soup has a diameter of 7.5 cm and a height of 11 cm. Calculate the volume of the can. **[2 marks]**

Radius = 3.75 cm
Volume = $\pi \times 3.75^2 \times 11 = 485.965...$
= 486 cm³ (3 s.f.)

KEY SKILLS
You need to be able to calculate the volume of a cylinder in context.

EXAM TIP
The formula for volume requires the radius of the cylinder.

WORKED EXAMPLE

A cylindrical can of soup has a diameter of 7.5 cm and a height of 11 cm.

Calculate:

a) the curved surface area of the can

b) the total surface area of the can. **[4 marks]**

a) Curved surface area $= \pi \times 7.5 \times 11$
$= 259.181...$
$= 259 \text{ cm}^2$ (3 s.f.)

b) Total surface area $=$ curved surface area $+ 2\pi r^2$
$= 259.181... + 2 \times \pi \times 3.75^2$
$= 347.538...$
$= 348 \text{ cm}^2$ (3 s.f.)

KEY SKILLS

You need to be able to calculate the surface area of a cylinder in context.

EXAM TIP

Use the given diameter to calculate the curved surface area.

Use the exact value of the curved surface area so as not to introduce rounding errors.

◀◀ RECAP

Pyramid

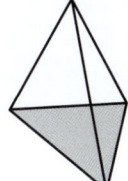 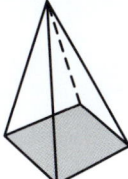

Volume $= \dfrac{1}{3} \times$ base area $\times$ height

Cone

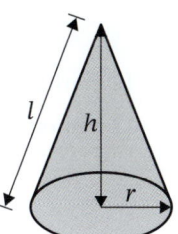

Volume $= \dfrac{1}{3}\pi r^2 h$

Curved surface area $= \pi r l$ where l is the 'slant height'

Sphere

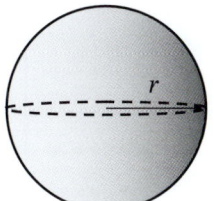

Volume $= \dfrac{4}{3}\pi r^3$

Surface area $= 4\pi r^2$

⊙ WATCH OUT!

There is no formula as such for the surface area of a pyramid.

It depends on, among other things, the shape and dimensions of the base.

147

5 Mensuration

WORKED EXAMPLE

A pyramid has a square base, with side length 8 cm, and a height of 12 cm.

Find the volume of the pyramid. [2 marks]

Volume = $\frac{1}{3} \times 8^2 \times 12 = 256 \text{ cm}^3$

WORKED EXAMPLE

A cone has base radius 6 cm and perpendicular height 8 cm.

Work out:

a) the exact volume of the cone [2 marks]

b) the exact curved surface area of the cone. [3 marks]

a) Volume = $\frac{1}{3} \pi \times 6^2 \times 8 = 96\pi \text{ cm}^3$

b) Curved surface area = $\pi r l$

Using Pythagoras' theorem:
$l^2 = r^2 + h^2 = 6^2 + 8^2$
$= 100$
Hence $l = 10$
Curved surface area = $\pi \times 6 \times 10 = 60\pi \text{ cm}^2$

WORKED EXAMPLE

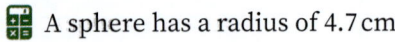 A sphere has a radius of 4.7 cm.

Find, correct to 3 significant figures:

a) the volume [2 marks]

b) the surface area. [2 marks]

a) Volume = $\frac{4}{3}\pi \times 4.7^3 = 434.89... = 435 \text{ cm}^3$ (3 s.f.)

b) Surface area = $4\pi \times 4.7^2 = 277.59... = 278 \text{ cm}^2$ (3 s.f.)

? QUESTIONS

1. A pyramid has a rectangular base measuring 5 cm by 6 cm.
   The perpendicular height of the pyramid is 8 cm.
   Calculate the volume of the pyramid.

2. A sphere has radius 6 cm.
   Find, giving your answers as exact multiples of π:
   a) the volume
   b) the surface area.

3. A cone has base radius 2.3 cm and perpendicular height 6.1 cm.
   Find, giving your answers to 3 significant figures:
   a) the volume
   b) the total surface area.

4. A cylindrical glass of radius 4 cm and height 9 cm is filled with water.
   The water is then poured into an upturned cone of base radius 5 cm and height 15 cm until the cone is full.
   How much water will be left in the glass?
   Give your answer to 2 significant figures.

5. A company makes spherical and cubical ice holders.
   The cubical container has a side length of 13 mm.
   The spherical container has the same volume as the cubical container.
   Calculate the radius of the spherical container.
   Give your answer in mm to 3 significant figures.

6. An ice cream scoop is designed to make spheres.
   Ice cream is taken from a container measuring 20 cm by 15 cm by 13 cm.
   The scoop always picks up perfect spheres of radius 2.4 cm.
   How many scoops (to the nearest whole number) can be filled from the container (assuming no waste)?

7. A 3-litre pot of paint is used to cover the surface of a large sphere.
   The instructions say that one litre of paint will cover 5 m^2.
   What is the maximum radius of the sphere if it is to be completely covered?

8. A cone of height 45 cm needs to hold at least 3 litres of water.
   What is the least possible value of the base radius?
   Give your answer to 3 significant figures.

9. The Earth is roughly a sphere with a radius of 6371 km.
   About 30% of the Earth's surface is land.
   What is this in km^2?
   Give your answer to 3 significant figures.

   To **Raise your grade** now try question 3 on page 153

5 Mensuration

5.5 Compound shapes

> **YOU NEED TO:**
> - Carry out calculations and solve problems involving perimeters and areas of compound shapes and parts of shapes.
> - Carry out calculations and solve problems involving the surface areas and volumes of compound shapes and parts of solids.

WORKED EXAMPLE

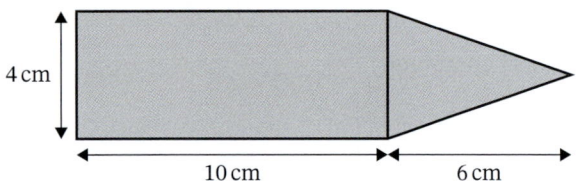

Calculate the area of this shape. **[3 marks]**

Area of rectangle = $4 \times 10 = 40 \text{ cm}^2$

Area of triangle = $\dfrac{4 \times 6}{2} = 12 \text{ cm}^2$

Total area = $40 + 12 = 52 \text{ cm}^2$

KEY SKILLS

You need to be able to find the area and/or perimeter of compound shapes.

EXAM TIP

Split the calculation into two parts and make sure your working is set out clearly.

WORKED EXAMPLE

An ice cream is modelled as a cone with a hemisphere on top.

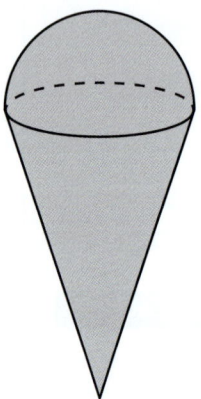

The height of the cone is 10 cm and the radius of the cone is 3 cm.

Find the total volume of the ice cream.

Give your answer as an exact multiple of π. **[4 marks]**

Volume of cone = $\dfrac{1}{3}\pi \times 3^2 \times 10 = 30\pi \text{ cm}^3$

Volume of hemisphere = $\dfrac{1}{2} \times \dfrac{4}{3}\pi \times 3^3 = 18\pi \text{ cm}^3$

Total volume = $30\pi + 18\pi = 48\pi \text{ cm}^3$

KEY SKILLS

You need to be able to work out the volume of a compound solid.

The formulas for the surface area and volume of a sphere, cone and pyramid will be given in the question when required.

You do not need to learn them.

EXAM TIP

Work out the two volumes separately.

Don't forget to halve the formula for the volume of a sphere.

5.5 Compound shapes

WORKED EXAMPLE

The cross-section of a water pipe is a circle of radius 10 cm.

Water flows along the pipe at a depth of 4 cm, as shown in the diagram.

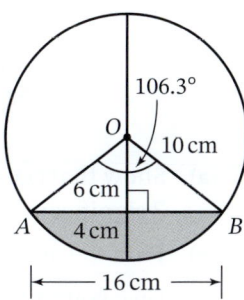

$OB = 10$ cm, $AB = 16$ cm and angle $AOB = 106.3°$

Given that the water is moving at 1.2 m per minute, calculate the volume of water that flows along the pipe in one hour.

Give your answer to 3 significant figures. **[6 marks]**

> The region shaded grey in the diagram is a segment of the circle.

> **WATCH OUT!**
> 6 marks indicate that this is going to be a multi-stage solution, so make sure that you set your working out neatly and logically.
>
> You might want to work in 'rough' first.

Area of cross-section of water = area of minor segment bounded by chord AB.

Area of minor segment = area of minor sector − area of triangle AOB.

Area of minor sector = $\dfrac{106.3}{360} \times \pi \times 10^2 = 92.764...$ cm^2

Area of triangle $AOB = \dfrac{6 \times 16}{2} = 48$ cm^2

Hence area of cross-section of water = $92.764... - 48 = 44.764...$ cm^2

Rate of flow = 120 cm per minute

Volume of water in 1 minute = $44.764... \times 120 = 5371.709...$ cm^3

So in 1 hour, volume of water = $5371.709... \times 60 = 322\,000$ cm^3 (3 s.f.)

> **EXAM TIP**
> Write down what you intend to calculate and how you are going to get there, then work the calculation through.
>
> Avoid using a rounded value for the cross-sectional area of the water. This ensures that your final answer will be correct when rounded.

? QUESTIONS

1. Find the area of this compound shape.

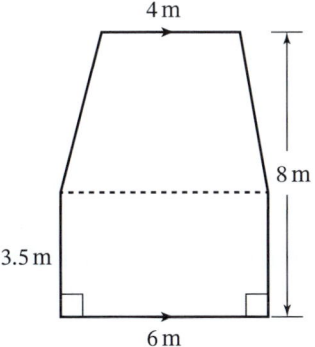

2. The area of this compound shape is 56 cm^2.

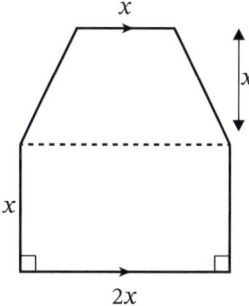

Find the value of x.

3. This shape consists of a square of side length 4 cm with a semicircle of radius 2 cm added to one side.

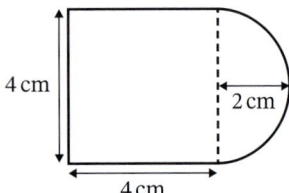

Show that the area of the shape is $16 + 2\pi$ cm².

4. A running track is 60 metres wide.

The inside of the inner lane consists of two straight sections of length 80 m and two semicircles of radius 24 m.

The outside of the outer lane consists of two straight sections of length 80 m and two semicircles of radius 30 m.

   a) Find, in terms of π, the perimeter of the inside of the inner lane.
   b) Find, in terms of π, the perimeter of the outside of the outer lane.
   c) Find, in terms of π, the area of the track.

5. A shape consists of a large semicircle with three smaller semicircles removed.

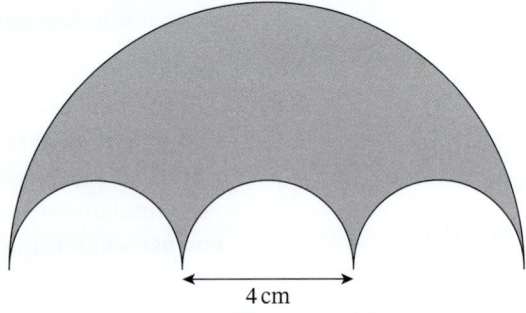

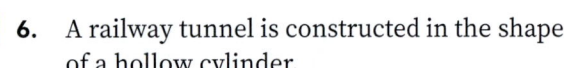

Each of the three smaller semicircles has diameter 4 cm.

Calculate the area of this shape. Leave your answer in terms of π.

6. A railway tunnel is constructed in the shape of a hollow cylinder.

It is 1 km long and has a radius of 3 m.

The base of the tunnel is covered with gravel to support the track.

A cross-section of the tunnel is shown in the diagram, with the shaded area representing the gravel.

AB represents the horizontal surface of the gravel.

X is the midpoint of AB and $\angle OAB = 60°$.

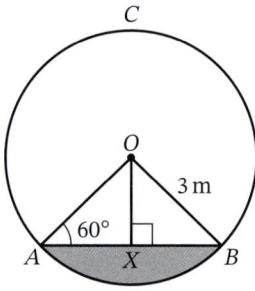

   a) Show that $OX = 2.60$ m, correct to 3 significant figures.
   b) Find the area of the triangle OAB.
      Give your answer to 3 significant figures.
   c) By considering the area of the sector OAB and your answer to part b), find the shaded area.
      Give your answer to 3 significant figures.
   d) Hence, find, correct to 3 significant figures, the volume of gravel required for the tunnel.
   e) Find the exact length of the major arc ACB.

The wall of the tunnel above the gravel level is to be painted.

   f) Find the surface area to be painted.
      Give an exact answer.

7. A ditch is cut in the ground in such a way that its cross-section is a trapezium, as shown in the diagram.

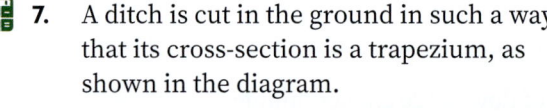

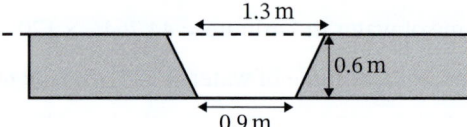

The ditch is 400 m long and it is filled with water.

   a) Calculate the area of the cross-section of the ditch.
   b) Calculate the volume of water contained in the ditch.

The water flows at a rate of 1.2 m s^{-1}.

   c) Calculate the volume of water that passes one point in a minute.

A pipe with a square cross-section is then placed in the ditch and the rest of the ditch is filled in with soil.

   d) If the pipe has the largest possible cross-sectional area, how much soil is put back into the ditch?

To **Raise your grade** now try questions 1, 2, 4 and 5 on page 153

↑ Raise your grade

1. Sam has a bowl made in the shape of an upside down frustum.

   The height of the bowl is 4 cm.

   Inside the bowl, the base has a radius of 2 cm and the top has a radius of 5 cm.

   a) Calculate the exact volume of the of the bowl. **[4 marks]**

   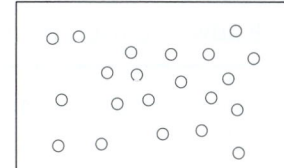

   Sam also has some metal spheres, each with a radius of 1 cm.

   100 cm³ of water is poured into the bowl.

   b) How many spheres can be placed into the bowl before the water starts to overflow? **[5 marks]**

2. Anne has a rectangular piece of paper.

   The paper has a length of 29.7 cm and a width of 21 cm.

   Anne starts punching little circles out of the paper using a hole punch that makes circles with a diameter of 6 mm.

   Assuming no holes overlap, how many holes can she punch out before she has less than half of the original piece of paper left? **[5 marks]**

3. Part of a parade float includes a number of papier mâchè cones, each with radius 4 cm and height 29 cm.

   The cones do not have a base.

   A small tin of paint can cover an area of 12 m².

   How many of these cones will one tin of paint cover completely? **[5 marks]**

4. A tin of soup has a diameter of 7.5 cm and a height of 11 cm.

   a) Show that the volume of the tin is 486 cm³, correct to 3 significant figures. **[2 marks]**

   The soup is poured from the tin into a bowl, whose shape is an upside down, hollow frustum.

   The diameter of the top of the bowl is 15 cm, the diameter of the base of the bowl is 8 cm, and the height of the bowl is also 8 cm.

   b) Assuming that the tin is completely full, will the contents fit in the bowl?
   Explain your reasoning. **[5 marks]**

5. In the diagram, the shape $AFBE$ is called a lune.

   The lune has two curved sides, AFB and AEB.

   $AFBO$ is a quarter circle of radius 10 cm, centred at O.

   AEB is a semicircle, radius AD, centred at D, where D is the midpoint of AB.

   Calculate the area of the lune. **[5 marks]**

   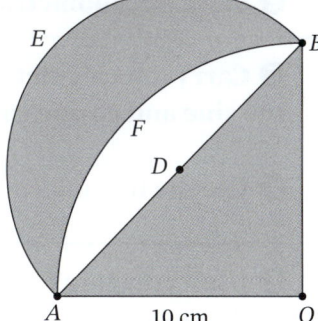

6. The diagram shows a sector of a circle.

   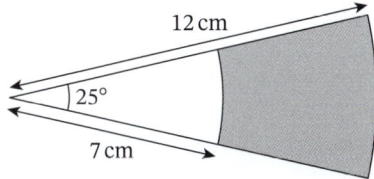

   Calculate:

   a) the area of the shaded region **[3 marks]**
   b) the perimeter of the shaded region. **[3 marks]**

7. The area of a right-angled triangle is 51.2 cm².

   The two perpendicular sides are in the ratio 1 : 1.6.

   Find the lengths of all three sides of the triangle. **[5 marks]**

6 Trigonometry

Your revision checklist

Tick these circles to build a record of your revision.

Core/ E Extended syllabus

6.1 Know and use Pythagoras' theorem.

6.2 Know and use the sine, cosine and tangent ratios for acute angles in calculations using sides and angles of a right-angled triangle.

Solve problems in two dimensions using Pythagoras' theorem and trigonometry.

E Know that the perpendicular distance from a point to a line is the shortest distance to the line.

E Solve problems involving angles of elevation and depression.

6.3 **E** Know the exact values of:
$\sin x$ and $\cos x$ for $x = 0°, 30°, 45°, 60°$ and $90°$, and

$\tan x$ for $x = 0°, 30°, 45°, 60°$.

6.4 **E** Recognise, sketch and interpret the graphs of $y = \sin x$, $y = \cos x$ and $y = \tan x$ for $0° \leq x \leq 360°$.

E Solve trigonometric equations involving $\sin x$, $\cos x$ or $\tan x$ for $0° \leq x \leq 360°$.

6.5 **E** Carry out calculations involving lengths and angles for any triangle using the sine and cosine rules.

E Use the formula: area of a triangle $= \dfrac{1}{2} ab \sin C$.

6.6 **E** Use Pythagoras' theorem and trigonometry to carry out calculations and solve problems in three dimensions, including calculating the angle between a line and a plane.

6.1, 6.2 Right-angled triangles

YOU NEED TO:

- Know and use Pythagoras' theorem.
- Know and use the sine, cosine and tangent ratios for acute angles in calculations involving sides and angles of a right-angled triangle.
- Solve problems in two dimensions using Pythagoras' theorem and trigonometry.
- **E** Know that the perpendicular distance from a point to a line is the shortest distance to the line.
- **E** Solve problems involving angles of elevation and depression.

◀◀ RECAP

Pythagoras' theorem states that:

'In a right-angled triangle the square on the hypotenuse is equal to the sum of the squares on the other two sides.'

$a^2 + b^2 = c^2$

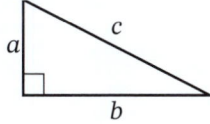

WORKED EXAMPLE

Find x, y and z in these triangles.

a)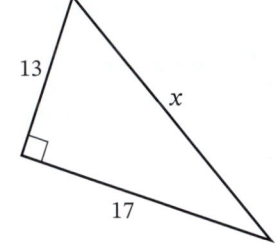

b)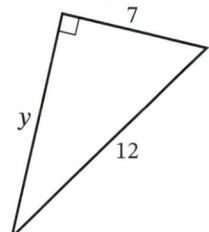

c)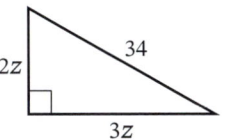

[7 marks]

a) $13^2 + 17^2 = x^2$
   $x^2 = 169 + 289 = 458$
   Hence $x = \sqrt{458} = 21.4$ (3 s.f.)

b) $y^2 + 7^2 = 12^2$
   $y^2 = 144 - 49 = 95$
   Hence $y = \sqrt{95} = 9.75$ (3 s.f.)

c) $(2z)^2 + (3z)^2 = 34^2$
   $4z^2 + 9z^2 = 1156$
   $13z^2 = 1156$
   $z = \sqrt{\dfrac{1156}{13}} = 9.43$ (3 s.f.)

🔑 KEY SKILLS

You need to be able to use Pythagoras' theorem to find a missing side length of a right-angled triangle.

👍 EXAM TIP

Show all of your working, even if you use a calculator.

For part **b)**, rearrange the formula to find the length of one of the shorter sides.

👍 EXAM TIP

The question may ask for an exact value, in which case leave your answer as a simplified surd.

👁 WATCH OUT!

$(2z)^2 = 4z^2$, not $2z^2$.

6 Trigonometry

⏪ RECAP

Remember SOH CAH TOA for right-angled triangles:

$$\sin \theta = \frac{\text{Opposite}}{\text{Hypotenuse}}$$

$$\cos \theta = \frac{\text{Adjacent}}{\text{Hypotenuse}}$$

$$\tan \theta = \frac{\text{Opposite}}{\text{Adjacent}}$$

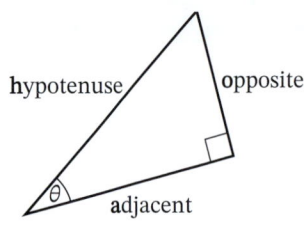

WORKED EXAMPLE

Find x and y in these triangles. [4 marks]

a)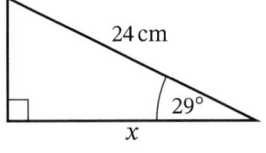

b)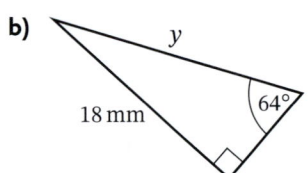

a) $\cos \theta = \dfrac{A}{H}$

$\cos 29° = \dfrac{x}{24}$

$x = 24 \times \cos 29°$

$\quad = 21.0 \text{ cm } (3 \text{ s.f.})$

b) $\sin \theta = \dfrac{O}{H}$

$\sin 64° = \dfrac{18}{y}$

$y = \dfrac{18}{\sin 64°}$

$\quad = 20.0 \text{ mm } (3 \text{ s.f.})$

🔑 KEY SKILLS

You need to be able to use trigonometry to find an unknown side length in a right-angled triangle.

👍 EXAM TIP

Write down the ratio you are going to use before substituting in.

In part **b)**, rearrange the equation carefully since your unknown is in the denominator of the fraction.

WORKED EXAMPLE

Find α in this triangle.

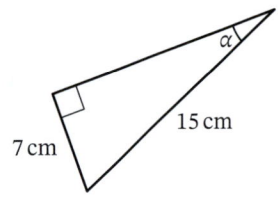

[2 marks]

$\sin \alpha = \dfrac{O}{H}$

$\sin \alpha = \dfrac{7}{15}$

$\alpha = \sin^{-1}\left(\dfrac{7}{15}\right)$

$\quad = 27.8° \ (1 \text{ d.p.})$

🔑 KEY SKILLS

You need to be able to use trigonometry to find an unknown angle in a right-angled triangle.

👍 EXAM TIP

Use the inverse trigonometric function on your calculator to find an angle.

Angles should be rounded to one decimal place unless otherwise stated.

Extended

⏪ RECAP

The **angle of elevation** is the angle you look up at to see something above you.

The **angle of depression** is the angle you look down at to see something below you.

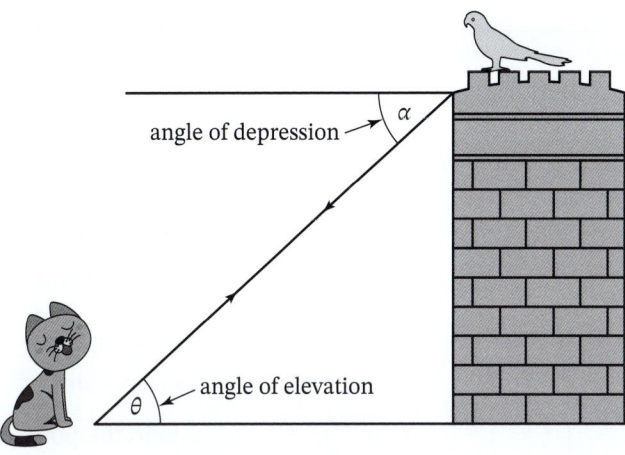

WORKED EXAMPLE

A surveyor, standing on the top of a building 30 m tall, sees two points C and D due north of her.

The angles of depression of C and D are 35° and 20°, respectively.

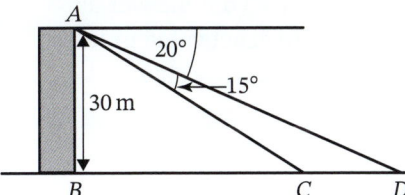

Find the distance CD. **[4 marks]**

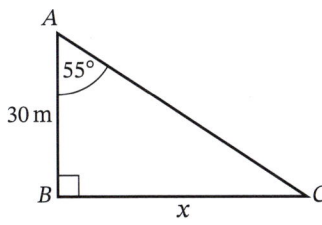   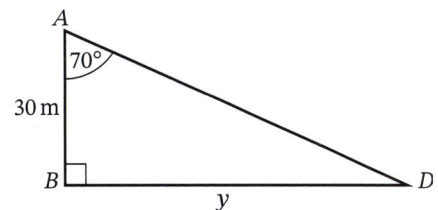

$\tan 55° = \dfrac{x}{30} \Rightarrow x = 30 \tan 55°$

$\tan 70° = \dfrac{y}{30} \Rightarrow y = 30 \tan 70°$

$CD = 30 \tan 70° - 30 \tan 55° = 39.6 \text{ m} \ (3 \text{ s.f.})$

👍 EXAM TIP

Use the information in the question to draw some clearly labelled diagrams.

The distance CD is $y - x$.

Work exactly and do not evaluate anything until the end. This prevents rounding errors creeping in.

6 Trigonometry

◀◀ RECAP

The shortest distance from a point to a line is the perpendicular distance.

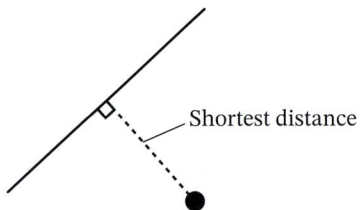

🔑 KEY SKILLS

You need to be able to find the shortest distance from a point to a line.

👁 WATCH OUT!

This problem is in the context of coordinate geometry so it is always best to draw a diagram.

WORKED EXAMPLE

Find the shortest distance from the point (3, 0) to the line with equation $y = \frac{1}{2}x + 1$ [5 marks]

$\frac{1}{2} \times m = -1 \Rightarrow m = -2$

So $y = -2x + c$

Using point (3, 0): $0 = -2 \times 3 + c \Rightarrow c = 6$

Hence the perpendicular line has equation $y = -2x + 6$

$\frac{1}{2}x + 1 = -2x + 6 \Rightarrow x = 2$

When $x = 2$, $y = 2$, so the point of intersection is (2, 2).

Distance $= \sqrt{(2-0)^2 + (2-3)^2}$
         $= \sqrt{5}$ units

👍 EXAM TIP

Find the equation of the line perpendicular to the given line through the given point.

Find the point of intersection of the two lines.

Use Pythagoras' theorem to find the distance between the given point and the point of intersection.

Give your answer as an exact value unless otherwise stated.

❓ QUESTIONS

1. A piece of A4 paper measures 298 mm by 210 mm.
   Find the length of the longest straight line that can be drawn on it.
   Give your answer to three significant figures.

2. A square field has side length 45 m.
   Find the length of the diagonal, giving your answer to 3 significant figures.

3. A person walks 1.5 km north and then 1 km east.
   How far are they from their starting point, to the nearest metre?

👍 EXAM TIP

Draw a diagram to show the situation.

158

4. An equilateral triangle has side length 8 cm.

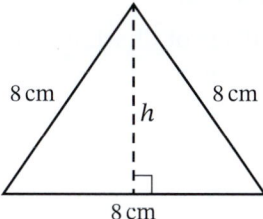

Find the exact height of the triangle.

5. Find the side lengths marked x in these triangles.

a)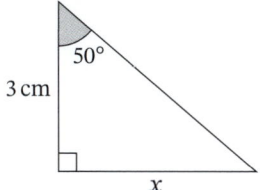

b)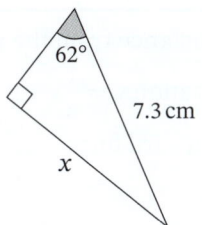

c)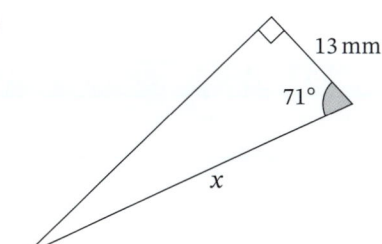

d)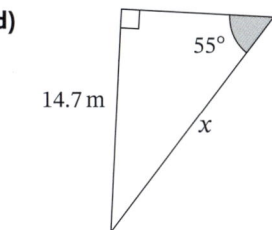

6. Find the angles marked x in these triangles.

a)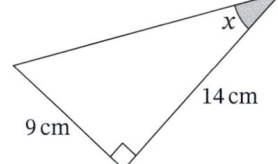

b)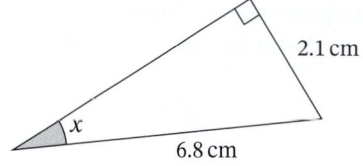

7. In the trapezium, $AD = 32$ cm, $FE = 20$ cm, $FB = 12$ cm and angle $FAB = 60°$.

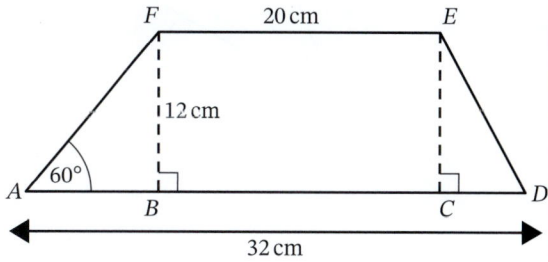

a) Show that $AB = 6.9$ cm (to 1 d.p.).
b) Calculate the size of angle EDC, giving your answer to one decimal place.

8. The diagram shows two right-angled triangles.

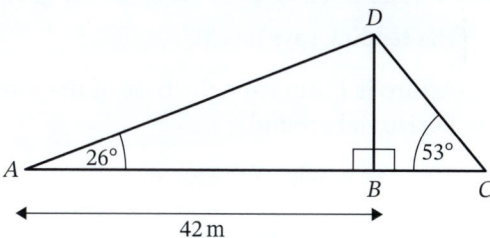

a) Show that the length BD is 20.5 m, correct to 3 significant figures.
b) Calculate the length DC, correct to 3 significant figures.
c) Find the size of angle ADC.

Give your answer to 1 decimal place.

9. In the diagram, D is the midpoint of AB.

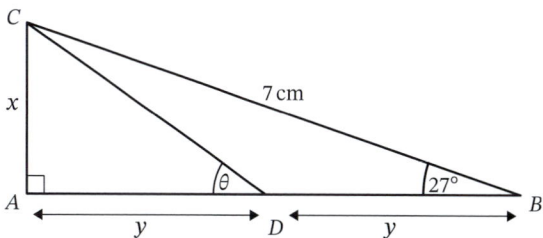

a) Find the length marked x, giving your answer to 3 significant figures.
b) Find the length marked y, giving your answer to 3 significant figures.
c) Hence, show that $\theta = 45.5°$, correct to 1 decimal place.

6 Trigonometry

10. The diagram shows three right-angled triangles.

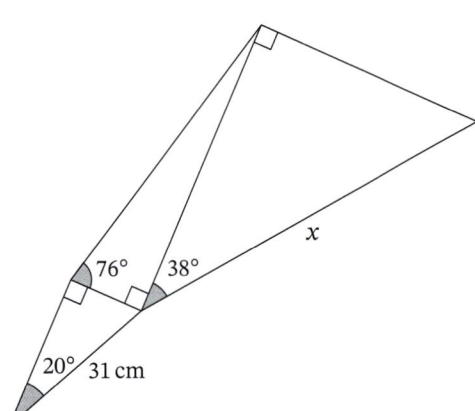

Find the length of the side marked x.

11. The town of Hinsdale is 30 km due south of Princeton.

The village of Charlbury is 25 km due east of Princeton.

Find the bearing of Hinsdale from Charlbury.

Extended

12. Arturo uses a clinometer to find the angle of elevation of a clock tower from the ground. His reading says it is 31°.

Arturo is 120 m from the base of the tower on horizontal ground.

Find the height of the tower.

13. Find the shortest distance from the point (2, 6) to the line with equation $y = \dfrac{1}{3}x + 2$.

Give your answer in surd form.

To **Raise your grade** now try questions 1, 2, 3 and 4 on page 171

6.3, 6.4 Trigonometric ratios and graphs

> **YOU NEED TO:**
>
> - **E** Know the exact values of sin x and cos x for $x = 0°, 30°, 45°, 60°$ and $90°$, and tan x for $x = 0°, 30°, 45°, 60°$.
> - **E** Recognise, sketch and interpret the graphs of $y = \sin x$, $y = \cos x$ and $y = \tan x$ for $0° \le x \le 360°$.
> - **E** Solve trigonometric equations involving sin x, cos x or tan x for $0° \le x \le 360°$.

Extended

◀◀ RECAP

You can find the sine, cosine or tangent of any angle.

You need to be able to work with angles in the interval $0° \le x \le 360°$.

You need to know the exact values of the sine, cosine or tangent of some angles.

x	0°	30°	45°	60°	90°
$\sin x$	0	$\dfrac{1}{2}$	$\dfrac{\sqrt{2}}{2}$	$\dfrac{\sqrt{3}}{2}$	1
$\cos x$	1	$\dfrac{\sqrt{3}}{2}$	$\dfrac{\sqrt{2}}{2}$	$\dfrac{1}{2}$	0
$\tan x$	0	$\dfrac{\sqrt{3}}{3}$	1	$\sqrt{3}$	–

The graph of $y = \sin x$ looks like this:

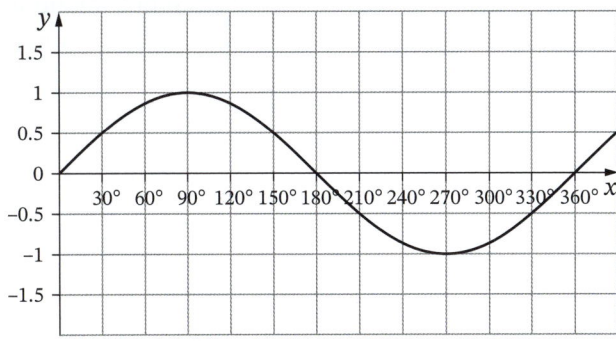

The graph of $y = \cos x$ looks like this:

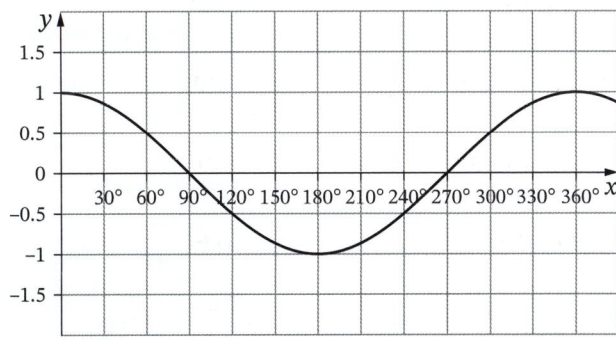

6 Trigonometry

The graph of $y = \tan x$ looks like this:

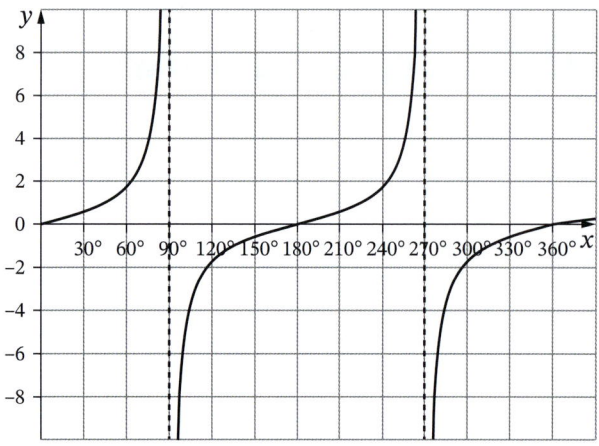

The dashed vertical lines drawn through $x = 90°$ and $x = 270°$ are called **asymptotes** and represent a discontinuity in the graph.

This is because $\tan x$ is undefined when $x = 90°$ or $x = 270°$.

WORKED EXAMPLE

Solve, in the interval $0° \leq x \leq 360°$, the following equations.

a) $\sin x = 0.5$

b) $2\tan x = 7$ [4 marks]

a) $\sin 30° = \dfrac{1}{2}$ so the first solution is $30°$

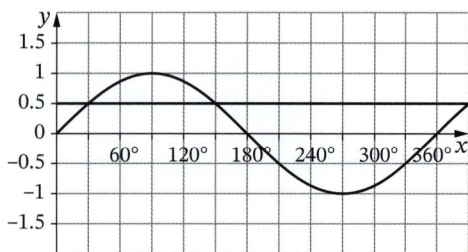

$\sin x = 0.5$ is also true when $x = 150°$

Hence, $x = 30°$ or $x = 150°$

b) $2\tan x = 7 \Rightarrow \tan x = 3.5$

$x = \tan^{-1}(3.5) = 74.054... = 74.1°$ (1 d.p.)

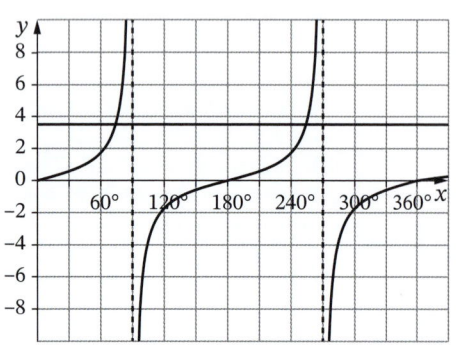

$\tan x = 3.5$ is also true when $x = 254.1°$

Hence, $x = 74.1°$ or $x = 254.1°$ (both to 1 d.p.)

🔑 KEY SKILLS

You need to be able to use the graphs of the trigonometric functions to solve trigonometric equations in the interval $0° \leq x \leq 360°$.

👍 EXAM TIP

In part **a)**, use your knowledge of the exact trigonometric values to find the **principal solution**.

Sketch a graph of the sine function and add a line going across at $y = 0.5$.

You can then read off other solutions to the equation and check them on your calculator.

👍 EXAM TIP

In part **b)**, rearrange the equation into the form $\tan x = ...$, then use the inverse tangent function to find the principal solution.

Again, sketch a graph.

Notice that the tan function repeats itself every 180° so you can find additional solutions accurately by adding/subtracting multiples of 180° to the principal solution.

6.3, 6.4 Trigonometric ratios and graphs

 QUESTIONS

1. Solve, in the interval $0° \leq x \leq 360°$, the following equations.
   a) $\sin x = 0.4$
   b) $\cos x = 0.5$
   c) $\tan x = 2$
   d) $\sin x = -0.6$
   e) $\cos x = -0.8$
   f) $\tan x = -1$

2. Solve, in the interval $0° \leq x \leq 360°$, the following equations.
   a) $2\sin x = 0.6$
   b) $3\cos x = -1$
   c) $5\tan x = 24$
   d) $1 + 3\sin x = 2$
   e) $2 - 3\cos x = 4$
   f) $6 - 5\tan x = -2$

To **Raise your grade** now try question 5 on page 171

6 Trigonometry

6.5 Non-right-angled triangles

YOU NEED TO:

- **E** Carry out calculations involving lengths and angles for any triangle using the sine and cosine rules.
- **E** Use the formula: area of a triangle = $\frac{1}{2}ab \sin C$.

Extended

◀◀ RECAP

The sine rule states that

$$\frac{a}{\sin A} = \frac{b}{\sin B} = \frac{c}{\sin C}$$

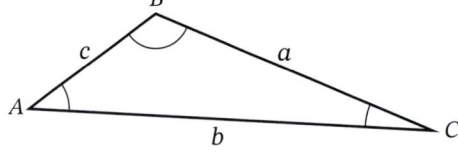

When finding an angle, rewrite the rule as

$$\frac{\sin A}{a} = \frac{\sin B}{b} = \frac{\sin C}{c}$$

Use the sine rule if you know one side and its opposite angle and one other measurement.

The cosine rule states that

$a^2 = b^2 + c^2 - 2bc \cos A$

Use the cosine rule if you know

- two sides and the included angle (the angle between the two sides), or
- all three sides.

🔑 KEY SKILLS

You need to be able to use the sine and cosine rules to find unknown side lengths or unknown angles in non-right-angled triangles.

WORKED EXAMPLE

 a) Find the length marked x. [2 marks]

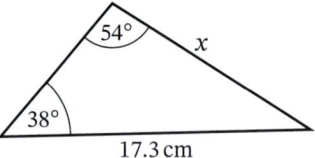

b) Find the size of the angle marked θ. [2 marks]

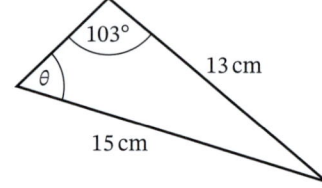

👍 EXAM TIP

In part **a)**, use the sine rule since you are working with two angles and two sides.

In part **b)**, use the alternative version of the sine rule since you are finding an angle.

a) $\dfrac{x}{\sin 38°} = \dfrac{17.3}{\sin 54°}$

$x = \dfrac{17.3}{\sin 54°} \times \sin 38°$

$= 13.165... = 13.2\,\text{cm}$ (3 s.f.)

b) $\dfrac{\sin \theta}{13} = \dfrac{\sin 103°}{15} \Rightarrow \sin \theta = \dfrac{\sin 103°}{15} \times 13 = 0.8444...$

Hence, $\theta = \sin^{-1}(0.8444...) = 57.61... = 57.6°$ (1 d.p.)

WORKED EXAMPLE

a) Find the length marked x. [2 marks]

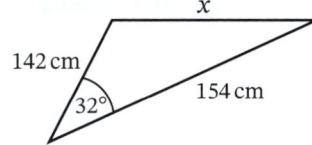

b) Find the size of the angle marked θ. [2 marks]

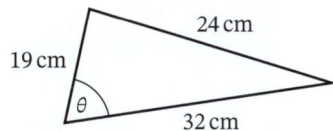

a) $x^2 = 142^2 + 154^2 - 2 \times 142 \times 154 \times \cos 32°$

$x^2 = 6789.768...$

Hence $x = 82.4\,\text{cm}$ (to 3 s.f.)

b) $24^2 = 32^2 + 19^2 - 2 \times 32 \times 19 \times \cos \theta$

$576 = 1385 - 1216 \cos \theta$

$\cos \theta = \dfrac{1385 - 576}{1216} = 0.6652...$

$\theta = \cos^{-1}(0.6652...) = 48.3°$ (1 d.p.)

👍 EXAM TIP

Use the cosine rule since you are working with all three sides and one angle.

In part **b)**, evaluate the numbers and then carefully rearrange the equation to make $\cos \theta$ the subject.

⏪ RECAP

Suppose you know the lengths of a, b and the angle C.

Area $= \dfrac{1}{2} ab \sin C$

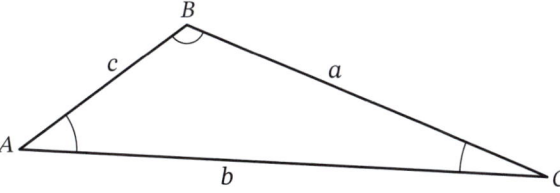

You can use this formula when you know the lengths of two sides in a triangle and the angle between them.

6 Trigonometry

WORKED EXAMPLE

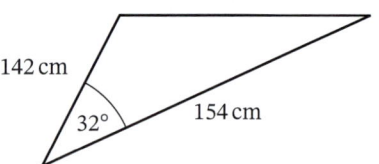

Find the area of the triangle. **[2 marks]**

$\frac{1}{2} \times 142 \times 154 \times \sin 32° = 5794.13...$

Area $= 5790 \text{ cm}^2$ (3 s.f.)

KEY SKILLS

You need to be able to calculate the area of a triangle using the general formula.

EXAM TIP

Use the general formula for the area of a triangle since you have two sides and the included angle.

WORKED EXAMPLE

 The diagram shows four cities, A, B, C and D.

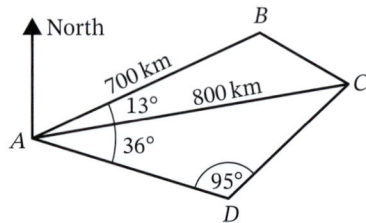

$AB = 700$ km and $AC = 800$ km.

Angle $BAC = 13°$, angle $CAD = 36°$ and angle $ADC = 95°$.

Calculate, giving your answers to three significant figures:

a) the distance BC
b) the distance AD
c) the area of land enclosed by the quadrilateral $ABCD$. **[8 marks]**

a) $BC^2 = 700^2 + 800^2 - 2 \times 700 \times 800 \times \cos 13°$
   $BC^2 = 38705.52...$
   Hence $BC = 196.73... = 197$ km (3 s.f.)

b) Angle $ACD = 180° - 95° - 36° = 49°$
   $\frac{AD}{\sin 49°} = \frac{800}{\sin 95°} \Rightarrow AD = \frac{800}{\sin 95°} \times \sin 49°$
   Hence $AD = 606.07... = 606$ km (3 s.f.)

c) Area of $ABCD$ = Area of triangle ABC + Area of triangle ACD
   Area of $ABC = \frac{1}{2} \times 700 \times 800 \times \sin 13° = 62986.29...$
   Area of $ACD = \frac{1}{2} \times 800 \times 606... \times \sin 36° = 142486.53...$
   Hence area of $ABCD = 205482.8... = 205000 \text{ km}^2$ (3 s.f.)

KEY SKILLS

You need to be able to solve practical problems using the sine and cosine rules and the general formula for the area of a triangle.

EXAM TIP

In part **a)**, use the cosine rule to find the third side of the triangle ABC.

In part **b)**, to use the sine rule in triangle ACD, you first need to find the angle opposite side AD.

Work as accurately as possible before rounding your final answer.

6.5 Non-right-angled triangles

QUESTIONS

1. Find the length of each side marked x.

   a)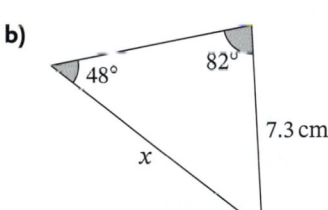

   b)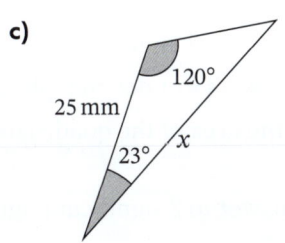

   c)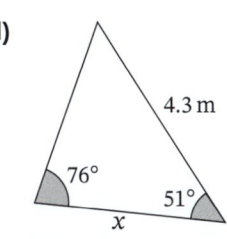

   d)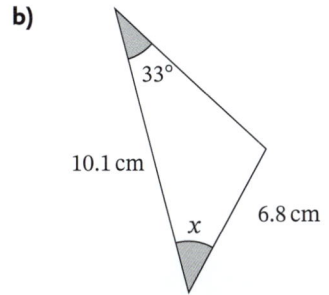

2. Find the size of each angle marked x.

   a)

   b)

3. A garden, ABC, is in the shape of a triangle.
   $AB = 7.8\,\text{m}$, angle $ACB = 65°$ and angle $BAC = 39°$.
   Find the length of BC.

4. An aircraft flies due north for 65 km and then turns onto a bearing of $x°$ for a further 80 km.
   The bearing of the aircraft from its original starting point is 065°.
   Find the value of x.

5. Find the length of each side marked x.

   a)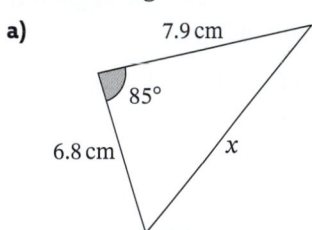

   b)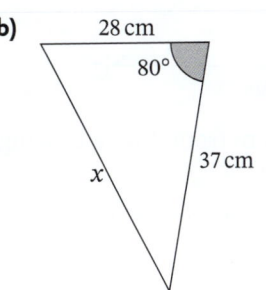

6. Find the size of each angle marked x.

   a)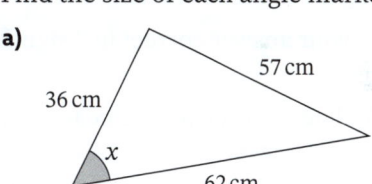

   b)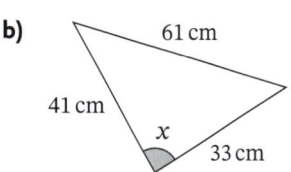

7. Hilda the hedgehog walks due north for 75 m and then turns through a bearing of 073° and walks a further 64 m.
   How far is she from her starting point?

167

8. The diagram shows the sail on a yacht.

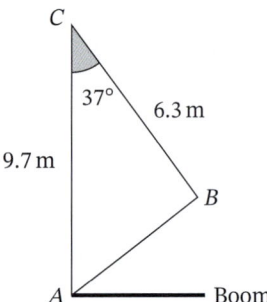

The mast AC is vertical.

Find the length of AB and the angle it makes with the horizontal boom.

9. Find the area of this triangle, correct to 2 decimal places.

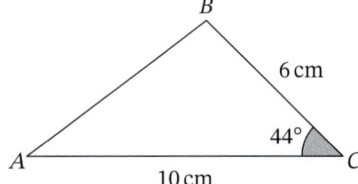

10. A car travels 15 km from A on a bearing of 100° to a point B.

It then travels 12 km from B on a bearing of 175° to a point.

The car then returns to A.

a) Calculate the area of triangle ABC

Give your answer correct to 3 significant figures.

b) Calculate the time (to the nearest minute) it takes to travel from C to A if the car maintains a steady speed of 80 km h^{-1}.

11. The diagram shows a quadrilateral ABCD.

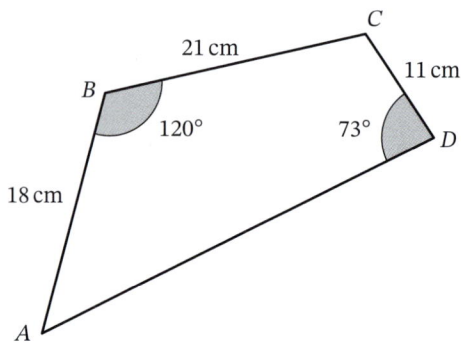

a) Find the length of AC.

Give your answer to 3 significant figures.

b) Find the area of triangle ABC.

Give your answer to 3 significant figures.

c) Find angle CAD and angle ACD

Give your answers to 1 decimal place.

d) Hence find the area of the quadrilateral ABCD.

Give your answer to 2 significant figures.

To **Raise your grade** now try questions 6, 7, 8 and 10 on pages 172–173

6.6 Trigonometry in 3D

> **YOU NEED TO:**
>
> - **E** Use Pythagoras' theorem and trigonometry to carry out calculations and solve problems in three dimensions, including calculating the angle between a line and a plane.

Extended

WORKED EXAMPLE

The diagram shows a model of an Egyptian pyramid.

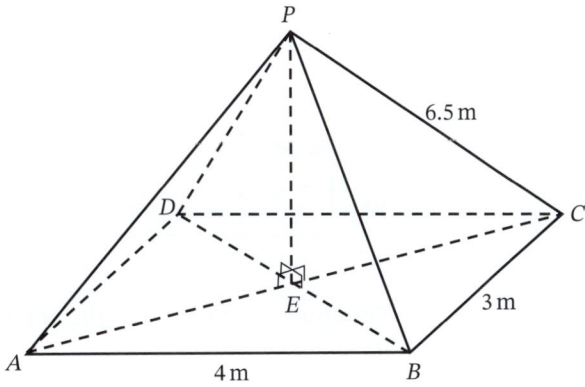

$AB = 4\,\text{m}$, $BC = 3\,\text{m}$ and $PC = 6.5\,\text{m}$.

$ABCD$ is a horizontal rectangular base and point E lies at the intersection of lines AC and BD.

P is vertically above E.

Calculate:

a) the height of the model

b) angle PCA

c) angle PBC. [9 marks]

a) Using Pythagoras' theorem on triangle ABC gives:

$x^2 = 3^2 + 4^2 = 25$

Hence, $x = 5\,\text{m}$ and $AE = 2.5\,\text{m}$

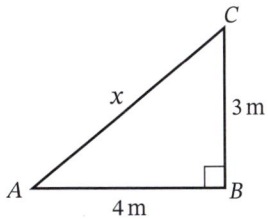

Using Pythagoras' theorem on triangle AEP gives:

$h^2 = 6.5^2 - 2.5^2 = 36$

Hence, $h = 6\,\text{m}$

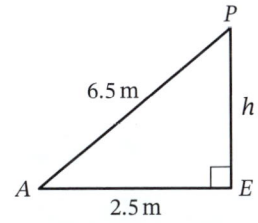

🔑 KEY SKILLS

You need to be able to find missing side lengths and unknown angles in 3D shapes.

👍 EXAM TIP

Break the problem down into stages and draw labelled diagrams of the triangles you are using.

6 Trigonometry

b) angle *PCA* = angle *PCE*

Using trigonometry:

$\cos\theta = \dfrac{2.5}{6.5}$; hence, $\theta = \cos^{-1}\left(\dfrac{2.5}{6.5}\right)$

Angle *PCA* = 67.4° (1 d.p.)

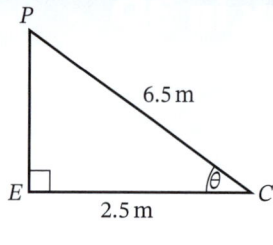

c) Angle *PBC* = Angle *PBM*, where *M* is the midpoint of *BC*.

$\cos\beta = \dfrac{1.5}{6.5}$ hence $\beta = \cos^{-1}\left(\dfrac{1.5}{6.5}\right)$

Angle *PBC* = 76.7° (1 d.p.)

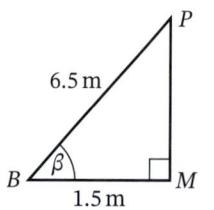

> **EXAM TIP**
>
> In part **b)**, look for right-angled triangles so you can use simple trigonometry.
>
> Do not forget to round angles to 1 decimal place.

QUESTIONS

1. Find the length of the longest diagonal in this cuboid.

   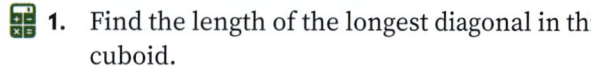

2. A cuboid has a longest diagonal equal to 75 cm.

   Two of the sides are 16 cm and 40 cm.

   Find the length of the third side.

3. A right square-based pyramid has a perpendicular height of 14 m and a base length of 12 m.

   Find the angle between the base and one of the sloping sides.

4. Find the height of this square-based pyramid.

   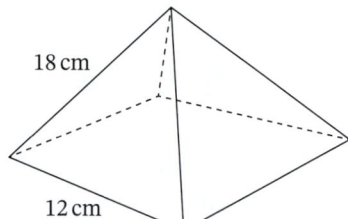

5. A mouse runs up this massive block of cheese, directly from point *A* to point *B*.

   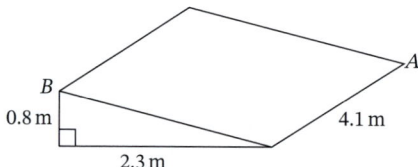

   Find how far it runs and the angle of the slope it runs up.

6. In the pyramid *ABCDE*, the square base *ABCD* is horizontal and *EM* is vertical.

   M is the midpoint of *AC*.

   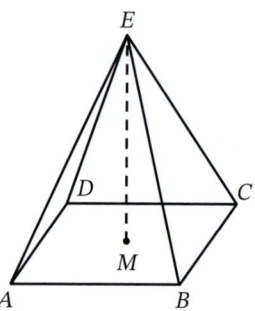

   AB = 15 cm and *EM* = 11 cm.

   a) Find *AM*.

      Give your answer to 3 significant figures.

   b) Show that *AE* = 15.3 cm, correct to 3 significant figures.

   c) Calculate the angle between the line *AE* and the plane *ABCD*.

      Give your answer to 1 decimal place.

   To **Raise your grade** now try questions 9 and 11 on pages 172–173

↑ Raise your grade

1. A circle is inscribed in a square as shown.

   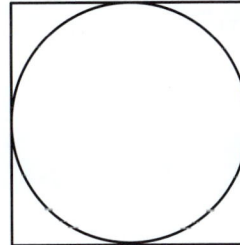

   The radius of the circle is 5 cm.
   Find the length of the diagonal of the square. [4 marks]

2. In the diagram below, AB is a tangent to the circle.

   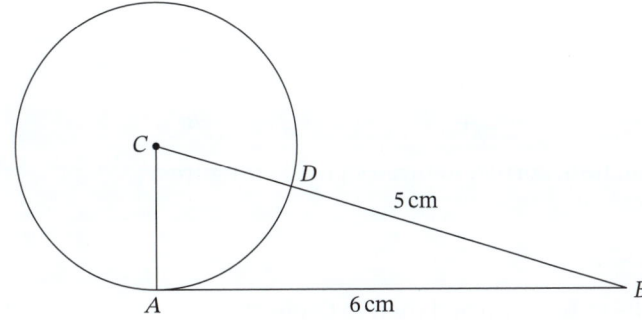

   AB = 6 cm and BD = 5 cm.
   Find the radius of the circle. [5 marks]

3. The diagram shows quadrilateral JKLM.

   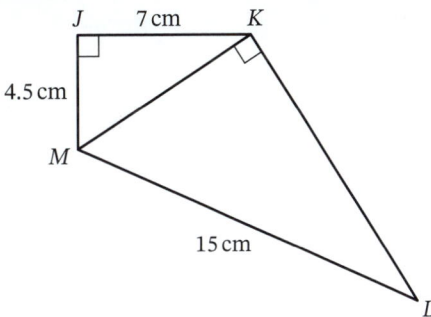

   Work out the size of angle KML.
   Give your answer to 1 decimal place. [4 marks]

Extended

4. Johnny is looking up at the top of a bell tower that is perpendicular to the ground.
   The angle of elevation of the bell tower is 35°, as measured from Johnny's feet.
   Johnny is standing 120 m from the tower.
   Find the height of the bell tower. [3 marks]

5. Given that $\sin x = \dfrac{\sqrt{2}}{2}$ and $0° \leq x \leq 180°$, find the two values of x that satisfy the equation. [3 marks]

6. In triangle ABC, $AB = 14\,cm$, $AC = 10\,cm$ and angle $CAB = 20°$.

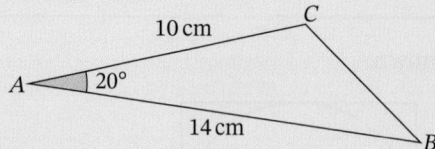

Calculate:

a) the length BC [2 marks]

b) the area of the triangle. [2 marks]

7. The triangle ABC has area $120\,cm^2$ and angle $ACB = 50°$.

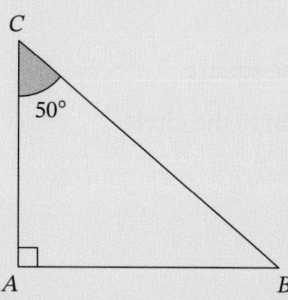

Given that $BC = 30\,cm$, find, correct to three significant figures:

a) AC [3 marks]

b) AB. [3 marks]

8. A boat sails from Aville to Beetown and then to Ceeford.

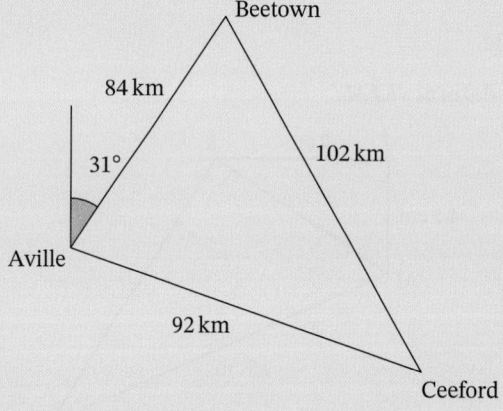

The bearing of Beetown from Aville is 031° and the distances are as shown in the diagram. Find the bearing of Aville from Ceeford. [6 marks]

9. A right square-based pyramid has base length $3.6\,cm$ and perpendicular height $5.1\,cm$.

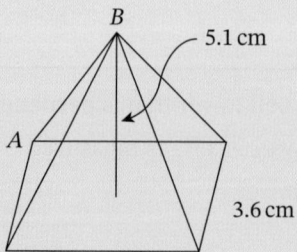

Find the length of edge AB. [6 marks]

10. The area of this triangle is 45 cm².

Find the length marked x. **[5 marks]**

11. Two cuboids are stacked as shown, with the top cuboid exactly in the centre of the bottom one.

Find the length marked PQ. **[6 marks]**

7 Transformations and vectors

Your revision checklist

Tick these circles to build a record of your revision.

Core/ E Extended syllabus

7.1 Recognise, describe and draw the following transformations:
- reflection of shapes in horizontal and vertical lines
- rotation of shapes in multiples of 90° about the origin
- enlargement of shapes from a centre by positive and fractional scale factors
- translation of shapes by a vector $\begin{pmatrix} x \\ y \end{pmatrix}$

- E reflection of shapes in horizontal, vertical and diagonal lines

- E rotation of shapes in multiples of 90° about the origin and other centres of rotation

- E enlargement of shapes from a centre by positive, fractional and negative scale factors

7.2 E Describe a translation using a vector represented by $\begin{pmatrix} x \\ y \end{pmatrix}$, $\overrightarrow{AB}$ or a.

E Add and subtract vectors.

E Multiply vectors by a scalar.

7.3 E Calculate the magnitude of a vector $\begin{pmatrix} x \\ y \end{pmatrix}$ as $\sqrt{x^2 + y^2}$.

7.4 E Represent vectors by directed line segments.

E Use position vectors.

E Express vectors in terms of two coplanar vectors using the sum and difference of two or more vectors.

E Use vectors to reason and to solve geometric problems.

174

7.1 Transformations

> **YOU NEED TO:**
> - Recognise, describe and draw the following transformations:
>   - reflection of shapes in horizontal and vertical lines
>   - rotation of shapes in multiples of 90° about the origin
>   - enlargement of shapes from a centre by positive and fractional scale factors
>   - translation of shapes by a vector $\begin{pmatrix} x \\ y \end{pmatrix}$
>   - **E** reflection of shapes in horizontal, vertical and diagonal lines
>   - **E** rotation of shapes in multiples of 90° about the origin and other centres of rotation
>   - **E** enlargement of shapes from a centre by positive, fractional and negative scale factors

> **◀◀ RECAP**
>
> You can transform a shape by reflecting (as in a mirror), rotating (turning), translating (moving) or enlarging it.
>
> The shape you start with is called the object and the shape you end with is called the image.

> **◀◀ RECAP**
>
> You describe a reflection by naming the mirror line.
>
> Triangle B is a reflection of triangle A in the line $x = 1$.
>
> Triangle A is the object.
>
> Triangle B is the image.

WORKED EXAMPLE

A triangle ABC has vertices at $A(-1, 1)$, $B(0, 1)$ and $C(1, 3)$.

Draw the image of triangle ABC when it is reflected in the line $y = x$, labelling the new vertices A', B' and C'. **[3 marks]**

First draw the triangle ABC.

Connect each vertex to the mirror line with a line perpendicular to the mirror line.

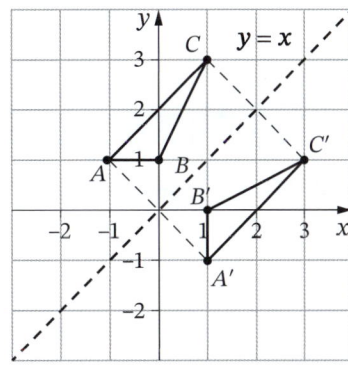

The distance from the line to the image is the same as the distance from the original point to the line.

> **🔑 KEY SKILLS**
>
> You must be able to reflect a simple plane figure and describe the transformation.

> **👍 EXAM TIP**
>
> At Core level, the mirror lines will be horizontal or vertical.

> **👍 EXAM TIP**
>
> Notice that when (p, q) is reflected in the line $y = x$ its image is (q, p).

7 Transformations and vectors

◀◀ RECAP

You describe a rotation by naming the centre, the angle and the direction (clockwise or anticlockwise) of the rotation.

The diagram shows an L-shape rotated through angles of 90°, 180° and 270° about a fixed point.

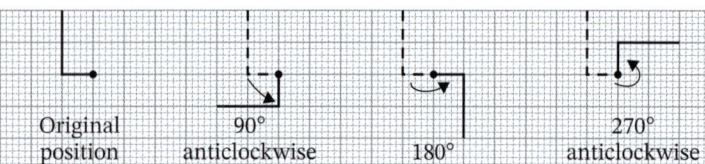

Original position | 90° anticlockwise | 180° | 270° anticlockwise

The fixed point is called the centre of rotation.

WORKED EXAMPLE

Find the image of the triangle with vertices $A(2, 2)$, $B(4, 5)$ and $C(6, 3)$ when it is rotated by 90° anticlockwise about the point $(0, 4)$.

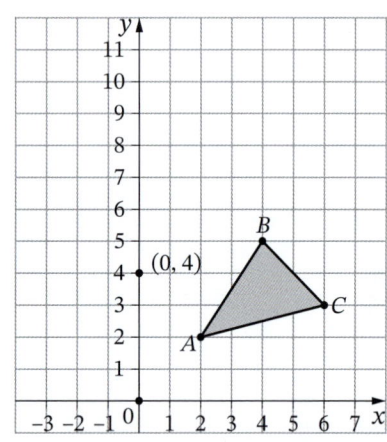

[3 marks]

🔑 KEY SKILLS

You must be able to rotate a simple plane figure and describe the transformation.

Imagine L-shapes joining $(0, 4)$ to the three vertices and then rotate these L-shapes as shown in the diagrams.

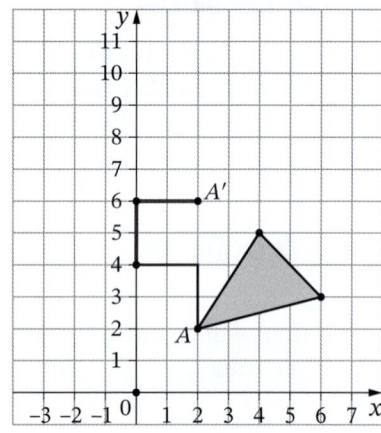

$A(2, 2)$ is mapped to $(2, 6)$

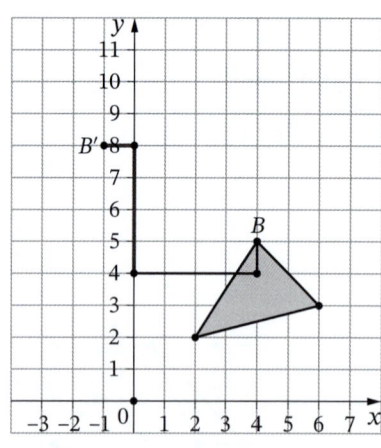

$B(4, 5)$ is mapped to $(-1, 8)$

👍 EXAM TIP

At Core level, the centre of rotation will be the origin, a vertex of the object, or a midpoint of a side of the object.

→

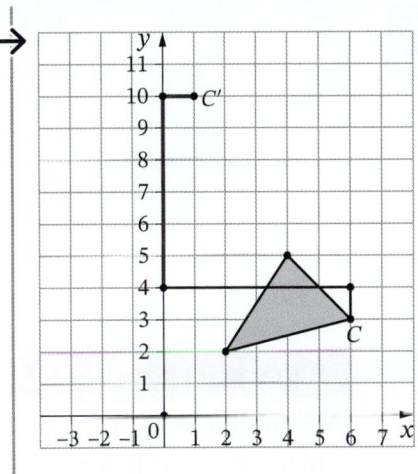

$C(6, 3)$ is mapped to $(1, 10)$

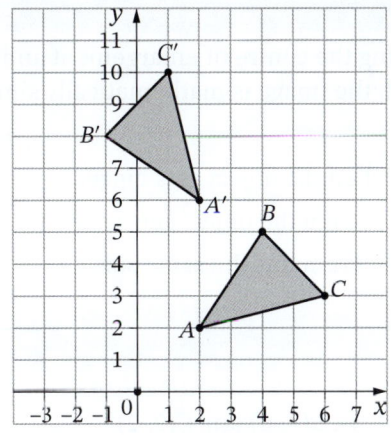

The image of ABC is then drawn to complete the rotation.

◀◀ RECAP

You describe a translation using a **vector**, $\begin{pmatrix} x \\ y \end{pmatrix}$.

- The top number in the vector tells you the horizontal movement (positive right, negative left).
- The bottom number in the vector tells you the vertical movement (positive up, negative down).

WORKED EXAMPLE

Translate triangle T using vector $\begin{pmatrix} 2 \\ 3 \end{pmatrix}$.

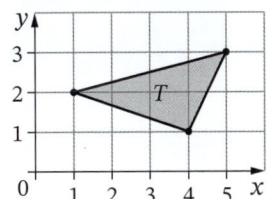

[2 marks]

Each corner of the triangle moves across to the right by 2 and up by 3.

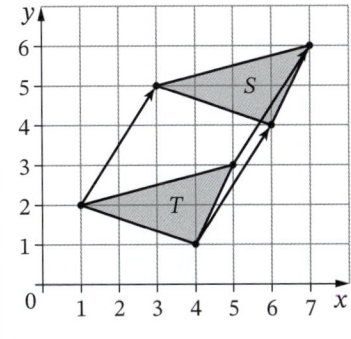

🗝 KEY SKILLS

You must be able to translate a simple plane figure and describe the transformation.

7 Transformations and vectors

◀◀ RECAP

You describe an enlargement by naming the centre of enlargement and the scale factor. When a shape is enlarged, the image is mathematically similar to the object.

- If the scale factor is greater than 1, then the object gets bigger.
- If the scale factor is between 0 and 1, then the object gets smaller.
- If the scale factor is 1, then the object remains the same size.

WORKED EXAMPLE

Enlarge triangle ABC with vertices $A(2, 4)$, $B(4, 0)$ and $C(2, 2)$ by scale factor 3 with centre of enlargement $X(1, 2)$. **[3 marks]**

🔑 KEY SKILLS

You must be able to enlarge a simple plane figure and describe the transformation.

To find A' draw the line XA extended through A.

Mark the point A' on this line such that $XA' = 3 \times XA$.

So A' is the point $(4, 8)$.

Repeat for B' and C'.

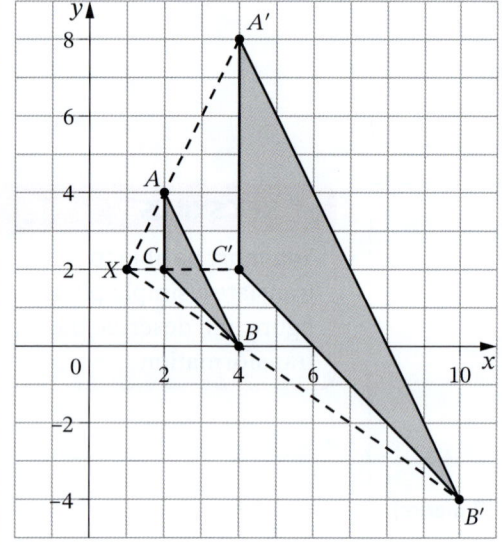

WORKED EXAMPLE

Enlarge quadrilateral $ABCD$ with vertices $A(-1, 1)$, $B(1, 3)$, $C(3, 3)$ and $D(3, -1)$ by scale factor $\frac{1}{2}$ with centre of enlargement $X(-3, -3)$. **[3 marks]**

To find A', draw the line XA.

Mark the point A' on this line such that

$XA' = \frac{1}{2} \times XA$

So A' is the point $(-2, -1)$

Repeat for B', C' and D'.

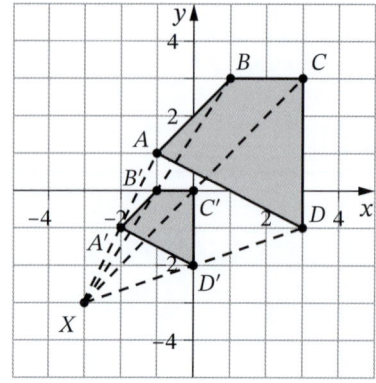

👁 WATCH OUT!

It is still referred to as an enlargement even when the image is smaller than the object.

7.1 Transformations

Extended

WORKED EXAMPLE

Enlarge quadrilateral $ABCD$ with vertices $A(2, 1)$, $B(4, 2)$, $C(4, 1)$ and $D(3, -1)$ by scale factor -2 with centre of enlargement $X(1, 0)$. **[3 marks]**

To find A' draw the line AX and extend past X.

Mark the point A' on this line such that $XA' = 2 \times XA$.

So A' is the point $(-1, -2)$.

Repeat for B', C' and D'.

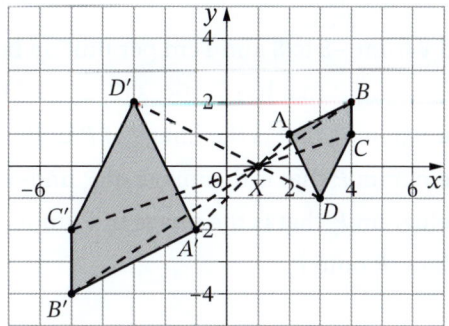

EXAM TIP

When the scale factor of enlargement is negative, the image is on the other side of the centre of enlargement from the object.

? QUESTIONS

1. Describe each of the transformations shown on the graph:

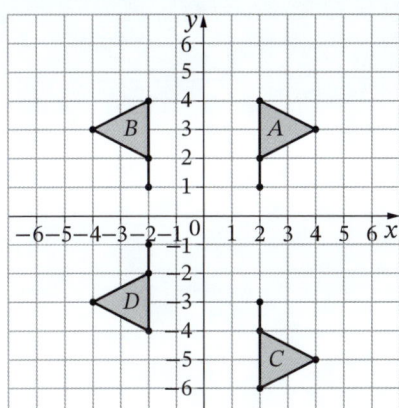

   a) A onto B b) A onto C c) A onto D

2. Describe each of the transformations shown on the graph:

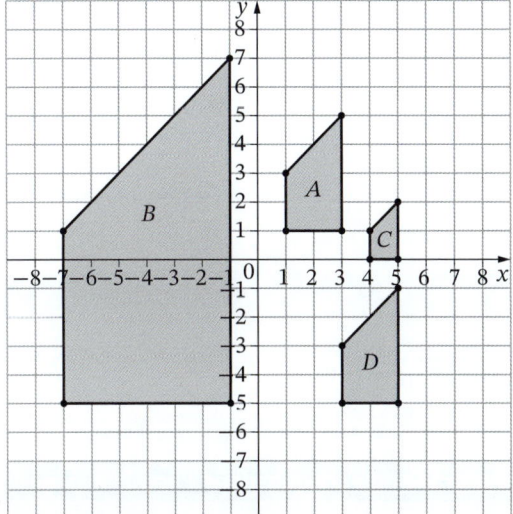

   a) A onto B b) A onto C c) A onto D
   d) D onto A e) C onto D

7 Transformations and vectors

3. Draw a pair of axes with x and y from -8 to 8 and $1\,\text{cm}$ per unit on both.
    a) Draw and label the triangle T_1 that has vertices $(2, 2)$, $(5, 2)$ and $(5, 4)$.
    b) Draw and label the triangle T_2 that is the image of T_1 under a rotation of $90°$ anticlockwise, centre $(2, 2)$.
    c) Draw and label the triangle T_3 that is the image of T_1 under a reflection in the line $y = 0$.
    d) Draw and label the triangle T_4 that is the image of T_1 under a translation using vector $\begin{pmatrix} -4 \\ -3 \end{pmatrix}$.
    e) Draw and label the triangle T_5 that is an enlargement of T_1, scale factor 2, centre $(6, 5)$.

4. Draw a pair of axes with x and y from -8 to 8 and $1\,\text{cm}$ per unit on both.
    a) Draw and label the parallelogram P_1 that has vertices $(2, 2)$, $(5, 2)$, $(3, 4)$ and $(6, 4)$.
    b) Draw and label the parallelogram P_2 that is the image of P_1 under a rotation of $180°$ about the origin.
    c) Draw and label the parallelogram P_3 that is the image of P_1 under a reflection in the line $x = 1$.
    d) Draw and label the parallelogram P_4 that is the image of P_1 under a reflection in the line $y = 0$.
    e) Find the vector that translates P_3 onto P_4.

Extended

5. Describe each of the transformations shown on the graph:

    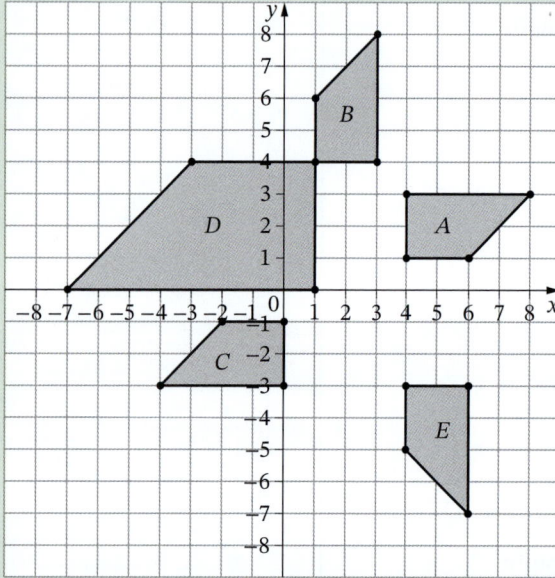

    a) A onto B
    b) A onto C
    c) A onto D
    d) A onto E.

6. Draw a pair of axes with x and y from -8 to 8 with $1\,\text{cm}$ per unit on both.
    a) Draw and label the triangle with vertices $A(1, 1)$, $B(2, 5)$ and $C(4, 3)$.
    b) Draw the rotation of triangle ABC by $90°$ anticlockwise, centre $(-1, 2)$.
        Label the image triangle $A_1B_1C_1$.
    c) Draw the reflection of triangle ABC in the line $y = -x$.
        Label the image triangle $A_2B_2C_2$.
    d) Draw the enlargement of triangle ABC, scale factor -2, with centre $(-1, 0)$.
        Label the image triangle $A_3B_3C_3$.

To **Raise your grade** now try questions 1 and 2 on page 186

7.2 Introduction to vectors

YOU NEED TO:

- **E** Describe a translation using a vector represented by $\begin{pmatrix} x \\ y \end{pmatrix}$, $\overrightarrow{AB}$ or **a**.
- **E** Add and subtract vectors.
- **E** Multiply vectors by a scalar.

Extended

⏪ RECAP

Look at this diagram.

The point $A(-3, 2)$ is translated to the point $B(-1, 5)$ by the column vector $\begin{pmatrix} 2 \\ 3 \end{pmatrix}$.

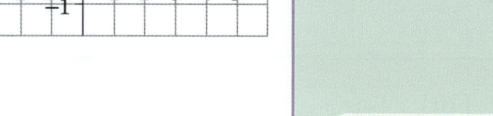

The notation $\overrightarrow{AB}$ is used to describe the translation from A to B.

Hence $\overrightarrow{AB} = \begin{pmatrix} 2 \\ 3 \end{pmatrix}$.

The same translation moves point C to point D.

Hence $\overrightarrow{CD} = \overrightarrow{AB}$.

A vector can also be represented by a lowercase letter. In textbooks, these are printed in **bold**. For example, $\mathbf{r} = \begin{pmatrix} 2 \\ 3 \end{pmatrix}$.

👍 EXAM TIP

Vectors that are set in bold font in print should be underlined when written by hand.

🗝 KEY SKILLS

You must be able to describe a translation using a vector in the three different formats.

WORKED EXAMPLE

The points A, B, C and D are $(2, 1)$, $(6, 3)$, $(1, 3)$ and $(5, 5)$ respectively.

a) Describe the vector from A to B as a column vector. **[1 mark]**
b) What can you say about the vectors $\overrightarrow{AB}$ and $\overrightarrow{CD}$? **[1 mark]**

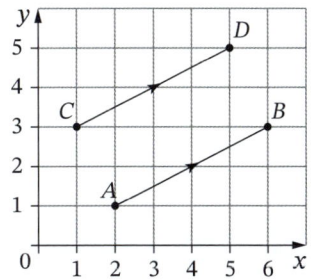

a) To get from A to B you go across by 4 and up by 2.

Hence $\overrightarrow{AB} = \begin{pmatrix} 4 \\ 2 \end{pmatrix}$.

b) They are equal.

👍 EXAM TIP

A vector is a directed line segment. If two vectors have the same magnitude and direction, then they are equal.

7 Transformations and vectors

WORKED EXAMPLE

$\mathbf{a} = \begin{pmatrix} 2 \\ 3 \end{pmatrix}$ and $\mathbf{b} = \begin{pmatrix} 4 \\ 1 \end{pmatrix}$

Write the following in column vector form.

a) $\mathbf{a} + \mathbf{b}$ [1 mark]

b) $\mathbf{a} - \mathbf{b}$ [1 mark]

a) $\mathbf{a} + \mathbf{b} = \begin{pmatrix} 2 \\ 3 \end{pmatrix} + \begin{pmatrix} 4 \\ 1 \end{pmatrix} = \begin{pmatrix} 6 \\ 4 \end{pmatrix}$

b) $\mathbf{a} - \mathbf{b} = \begin{pmatrix} 2 \\ 3 \end{pmatrix} - \begin{pmatrix} 4 \\ 1 \end{pmatrix} = \begin{pmatrix} -2 \\ 2 \end{pmatrix}$

KEY SKILLS
You must be able to add and subtract column vectors.

RECAP
To add or subtract vectors you just add or subtract the corresponding horizontal and vertical components.

WORKED EXAMPLE

$\mathbf{a} = \begin{pmatrix} 2 \\ 3 \end{pmatrix}$ and $\mathbf{b} = \begin{pmatrix} 4 \\ 1 \end{pmatrix}$

Write the following in column vector form.

a) $5\mathbf{a}$ [1 mark]

b) $3\mathbf{a} + 2\mathbf{b}$ [1 mark]

a) $5\mathbf{a} = 5\begin{pmatrix} 2 \\ 3 \end{pmatrix} = \begin{pmatrix} 10 \\ 15 \end{pmatrix}$

b) $3\mathbf{a} + 2\mathbf{b} = 3\begin{pmatrix} 2 \\ 3 \end{pmatrix} + 2\begin{pmatrix} 4 \\ 1 \end{pmatrix} = \begin{pmatrix} 14 \\ 11 \end{pmatrix}$

KEY SKILLS
You must be able to multiply a vector by a scalar.

RECAP
To multiply a vector by a scalar, multiply each component of the vector by the scalar.

EXAM TIP
Remember that a scalar just means a number.

QUESTIONS

1. $\mathbf{a} = \begin{pmatrix} 1 \\ 3 \end{pmatrix}$, $\mathbf{b} = \begin{pmatrix} -2 \\ 5 \end{pmatrix}$ and $\mathbf{c} = \begin{pmatrix} -4 \\ -1 \end{pmatrix}$.

   Write each of the following in column vector form.

   a) $\mathbf{a} + \mathbf{b}$
   b) $3\mathbf{a}$
   c) $-\mathbf{b}$
   d) $2\mathbf{b} - \mathbf{a}$
   e) $2\mathbf{a} - \mathbf{c}$
   f) $-\mathbf{a} - \mathbf{c}$
   g) $\mathbf{a} - 2\mathbf{b} + \mathbf{c}$
   h) $2(\mathbf{a} + \mathbf{b} - \mathbf{c})$
   i) $-4\mathbf{c}$
   j) $2\mathbf{c} - 3\mathbf{a}$
   k) $3\mathbf{a} - 2\mathbf{b}$
   l) $-2\mathbf{a} - \mathbf{b}$
   m) $-3\mathbf{c} + \mathbf{b}$
   n) $\frac{1}{2}(\mathbf{b} - \mathbf{c})$
   o) $\frac{1}{2}(\mathbf{c} - \mathbf{b})$

To **Raise your grade** now try question 3 on page 187

7.3, 7.4 Solving problems with vectors

YOU NEED TO:

- **E** Calculate the magnitude of a vector $\begin{pmatrix} x \\ y \end{pmatrix}$ as $\sqrt{x^2 + y^2}$.
- **E** Represent vectors by directed line segments.
- **E** Use position vectors.
- **E** Express vectors in terms of two coplanar vectors using the sum and difference of two or more vectors.
- **E** Use vectors to reason and to solve geometric problems.

Extended

WORKED EXAMPLE

$\mathbf{a} = \begin{pmatrix} 3 \\ 2 \end{pmatrix}$ and $\mathbf{b} = \begin{pmatrix} 7 \\ 5 \end{pmatrix}$

Find the magnitude of $\mathbf{a} + \mathbf{b}$. **[2 marks]**

$\mathbf{a} + \mathbf{b} = \begin{pmatrix} 3 \\ 2 \end{pmatrix} + \begin{pmatrix} 7 \\ 5 \end{pmatrix} = \begin{pmatrix} 10 \\ 7 \end{pmatrix}$

$|\mathbf{a} + \mathbf{b}| = \sqrt{10^2 + 7^2} = \sqrt{149} = 12.2$ (3 s.f.)

WORKED EXAMPLE

The diagram shows three vectors.

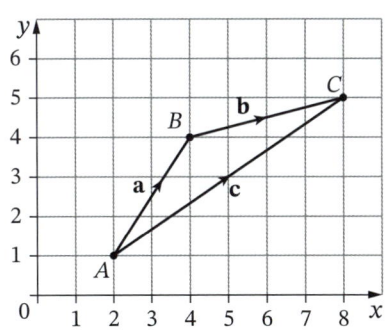

Express the vector $\mathbf{c}$ in terms of the two coplanar vectors $\mathbf{a}$ and $\mathbf{b}$. **[1 mark]**

The vector $\mathbf{c}$ takes you from point A to point C.

Another way to get from point A to point C is to go via point B.

This is vector $\mathbf{a}$ followed by vector $\mathbf{b}$.

Therefore $\mathbf{c} = \mathbf{a} + \mathbf{b}$.

KEY SKILLS

You must be able to calculate the magnitude of a vector.

RECAP

The magnitude of a vector $\mathbf{a} = \begin{pmatrix} x \\ y \end{pmatrix}$ is calculated using

$|\mathbf{a}| = \sqrt{x^2 + y^2}$

where $|\mathbf{a}|$ is the symbol for the magnitude of vector $\mathbf{a}$.

KEY SKILLS

You must be able to use the sum or difference of two coplanar vectors to express another given vector.

EXAM TIP

The word 'coplanar' just means 'in the same two-dimensional plane'. When you are working with vectors in two dimensions, all vectors are coplanar so you don't have to worry about this.

7 Transformations and vectors

◀◀ RECAP

A position vector is a vector that begins at a fixed origin, usually the point (0, 0) which is labelled O.

The position vector of point A is $\overrightarrow{OA} = \begin{pmatrix} 2 \\ 4 \end{pmatrix}$.

The position vector of point B is $\overrightarrow{OB} = \begin{pmatrix} 5 \\ 5 \end{pmatrix}$.

The position vector of point C is $\overrightarrow{OC} = \begin{pmatrix} 6 \\ 1 \end{pmatrix}$.

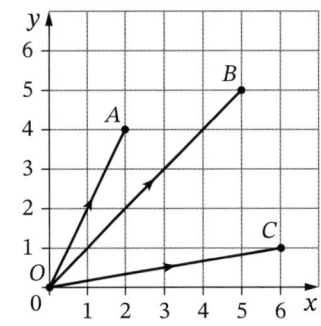

✎ APPLY

Look again at the vectors in the Worked example on the previous page.

How would you express vector **a** in terms of the vectors **b** and **c**?

How would you express vector **b** in terms of the vectors **a** and **c**?

WORKED EXAMPLE

Points A and B have position vectors $\overrightarrow{OA} = \begin{pmatrix} 3 \\ 4 \end{pmatrix}$ and $\overrightarrow{OB} = \begin{pmatrix} 9 \\ 5 \end{pmatrix}$ respectively.

Write the vector $\overrightarrow{AB}$ as column vector. **[2 marks]**

$\overrightarrow{AB} = \overrightarrow{OB} - \overrightarrow{OA} = \begin{pmatrix} 9 \\ 5 \end{pmatrix} - \begin{pmatrix} 3 \\ 4 \end{pmatrix} = \begin{pmatrix} 6 \\ 1 \end{pmatrix}$

⚷ KEY SKILLS

You must be able to use position vectors.

◀◀ RECAP

A useful formula for position vectors is

$\overrightarrow{AB} = \overrightarrow{OB} - \overrightarrow{OA}$

? QUESTIONS

1. Calculate the magnitude of each vector. Give exact answers.

   a) $\begin{pmatrix} 7 \\ 2 \end{pmatrix}$ b) $\begin{pmatrix} 1 \\ 14 \end{pmatrix}$ c) $\begin{pmatrix} 2 \\ 9 \end{pmatrix}$

2. Calculate the magnitude of each vector. Give your answers correct to 3 significant figures.

   a) $\begin{pmatrix} 1 \\ 5 \end{pmatrix}$ b) $\begin{pmatrix} 6 \\ -7 \end{pmatrix}$ c) $\begin{pmatrix} -8 \\ -12 \end{pmatrix}$

3. ABCD is a parallelogram such that $\overrightarrow{AB} = \mathbf{p}$ and $\overrightarrow{BC} = \mathbf{q}$.

M is the midpoint of AB and N is the midpoint of AC.

Point P is such that the ratio $AP : PC = 2 : 1$

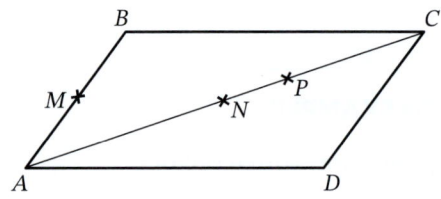

Find, in terms of **p** and **q**:

a) $\overrightarrow{CD}$ b) $\overrightarrow{AD}$ c) $\overrightarrow{AC}$

d) $\overrightarrow{AM}$ e) $\overrightarrow{AN}$ f) $\overrightarrow{AP}$

4. *ABCD* is a parallelogram with $\vec{AB} = \mathbf{p}$ and $\vec{AD} = \mathbf{q}$.

   M is the midpoint of *AD* and *N* is the midpoint of *AB*.

   R is the point one-quarter of the way from *A* to *C*.

   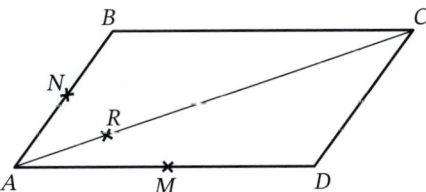

   a) Find, in terms of **p** and **q**:
   - i) $\vec{AM}$
   - ii) $\vec{AN}$
   - iii) $\vec{AC}$
   - iv) $\vec{AR}$
   - v) $\vec{MR}$
   - vi) $\vec{RN}$

   b) Hence, show that *M*, *N* and *R* all lie on a straight line.

   c) Find the ratio *MR* : *MN*.

5. *ABC* is a triangle with $\vec{AB} = \mathbf{p}$ and $\vec{AC} = \mathbf{q}$.

   X, *Y* and *Z* are the midpoints of *AB*, *BC* and *CA*, respectively.

   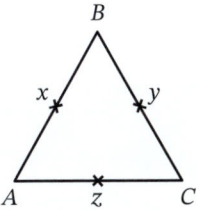

   a) Find, in terms of **p** and **q**:
   - i) $\vec{BC}$
   - ii) $\vec{BY}$
   - iii) $\vec{XB}$
   - iv) $\vec{XY}$
   - v) $\vec{YC}$
   - vi) $\vec{YZ}$
   - vii) $\vec{AZ}$
   - viii) $\vec{XZ}$

   b) Hence, show that *XZ* is parallel to *BC*.

   c) Find the ratio *XZ* : *BC*.

6. *ABCD* is a parallelogram with $\vec{AB} = \mathbf{r}$ and $\vec{AD} = \mathbf{s}$.

   X is two-thirds of the way along *BD* from *B* and *Y* is one-third of the way along *AD* from *A*.

   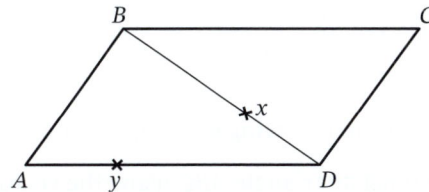

   a) Find, in terms of **r** and **s**:
   - i) $\vec{BD}$
   - ii) $\vec{AC}$
   - iii) $\vec{BX}$
   - iv) $\vec{AX}$
   - v) $\vec{AY}$
   - vi) $\vec{YX}$

   b) Hence show that *YX* is parallel to *AC*.

   c) Find the ratio *YX* : *AC*.

To **Raise your grade** now try questions 4, 5 and 6 on page 187

7 Transformations and vectors

↑ Raise your grade

1. Consider the triangle *ABC* shown below.

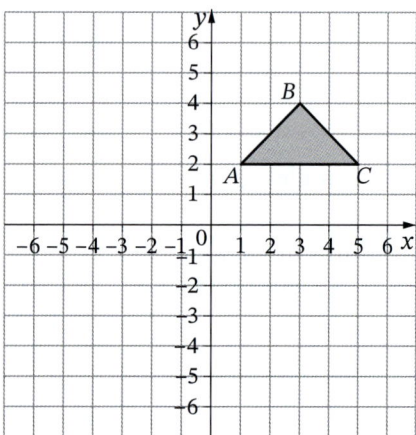

Answer the following on a copy of the diagram.

a) Translate triangle *ABC* using the vector $\begin{pmatrix} -5 \\ -4 \end{pmatrix}$ and label the image Δ2. [2 marks]

b) Reflect Δ2 in the line $y = 1$ and label this image Δ3. [2 marks]

c) Rotate Δ3 by 180° about (0, 0) and label this image Δ4. [2 marks]

d) Describe fully the single transformation that maps *ABC* onto Δ4. [2 marks]

Extended

2. The diagram shows shape *A*.

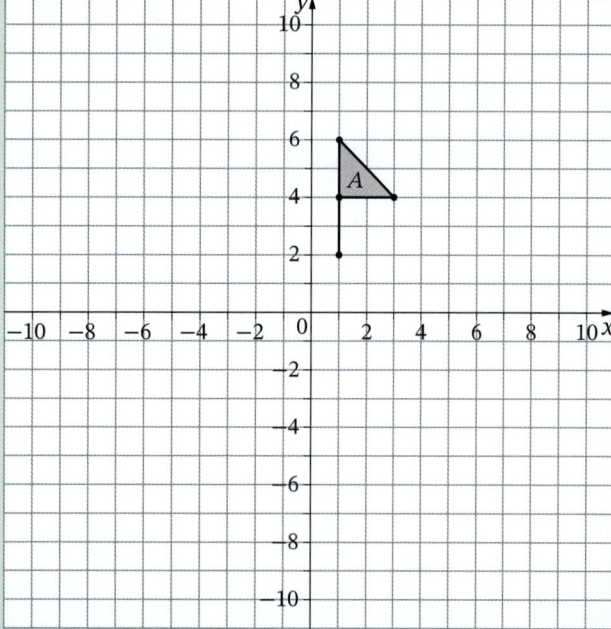

Answer the following on a copy of the diagram.

a) Reflect shape *A* in the line $y = x$ and label the image *B*. [2 marks]

b) Rotate shape *A* through 90° anticlockwise about (1, 1) and label the image *C*. [2 marks]

c) Enlarge shape *A* using scale factor −2 and centre of enlargement (−1, 1).
   Label the image *D*. [2 marks]

d) Describe fully the single transformation that maps shape *B* onto shape *C*. [2 marks]

3. $\mathbf{a} = \begin{pmatrix} 1 \\ 3 \end{pmatrix}$, $\mathbf{b} = \begin{pmatrix} -2 \\ 5 \end{pmatrix}$ and $\mathbf{c} = \begin{pmatrix} -4 \\ -1 \end{pmatrix}$

   Write each of these as a single column vector.

   a) $-7\mathbf{c}$ [2 marks]

   b) $\frac{1}{2}\mathbf{a}$ [2 marks]

   c) $-\frac{1}{2}(\mathbf{b} + \mathbf{c})$ [2 marks]

4. Calculate the magnitudes of the following vectors.

   Give your answers correct to 3 significant figures, where necessary.

   a) $\begin{pmatrix} -7 \\ 15 \end{pmatrix}$ [2 marks]

   b) $\begin{pmatrix} 11 \\ -4 \end{pmatrix}$ [2 marks]

   c) $\begin{pmatrix} 0 \\ -25 \end{pmatrix}$ [1 mark]

5. In the diagram, $\overrightarrow{OA} = \mathbf{a}$ and $\overrightarrow{OB} = \mathbf{b}$.

   M is the midpoint of the line segment AB.

   C is the midpoint of segment OA.

   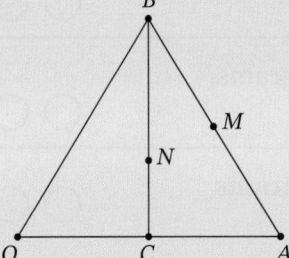

   The point N is such that $CN : NB = 1 : 2$.

   a) Find, in terms of $\mathbf{a}$ and $\mathbf{b}$:

   i) $\overrightarrow{AB}$ ii) $\overrightarrow{ON}$ [4 marks]

   b) Line ON is extended so that it passes through the line segment AB.

   Prove that this line passes through point M. [2 marks]

6. In the diagram, $\overrightarrow{AB} = \mathbf{a}$ and $\overrightarrow{BM} = \mathbf{b}$.

   M is the midpoint of $\overrightarrow{BC}$.

   The point N is such that $AN : NC = 2 : 1$.

   a) Express $\overrightarrow{AN}$ in terms of $\mathbf{a}$ and $\mathbf{b}$. [2 marks]

   b) Show that $\overrightarrow{NM} = \frac{1}{3}(\mathbf{a} - \mathbf{b})$. [3 marks]

   c) Show that $\overrightarrow{BN} = \frac{1}{3}(4\mathbf{b} - \mathbf{a})$. [3 marks]

   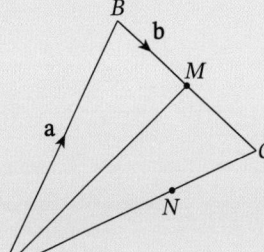

8 Probability

Your revision checklist

Tick these circles to build a record of your revision.

Core/ **E** **Extended** syllabus

		😟 😐 🙂
8.1	Understand and use the probability scale from 0 to 1.	○ ○ ○
	E **Understand and use probability notation.**	○ ○ ○
	Calculate the probability of a single event.	○ ○ ○
	Understand that the probability of an event not occurring is 1 minus the probability of the event occurring.	○ ○ ○
8.2	Understand relative frequency as an estimate of probability.	○ ○ ○
	Calculate expected frequencies.	○ ○ ○
8.3	Calculate the probability of combined events using sample space diagrams, Venn diagrams and tree diagrams, as appropriate.	○ ○ ○
8.4	**E** **Calculate conditional probability using Venn diagrams, tree diagrams and tables.**	○ ○ ○

8.1 Single events

YOU NEED TO:
- Understand and use the probability scale from 0 to 1.
- ⒺUnderstand and use probability notation.
- Calculate the probability of a single event.
- Understand that the probability of an event not occurring is 1 minus the probability of the event occurring.

◀◀ RECAP

The probability of an event is a number between 0 and 1 that tells you how likely it is that the event will occur.

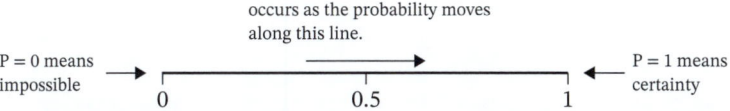

👍 EXAM TIP
In real life, not many things actually have probabilities of 0 or 1, because not many things are totally impossible (0) or completely certain (1).

Examples of probability

The probability of Al Ahly FC winning their next game is 0.75.

The probability of a part made in a factory being faulty is $\frac{1}{50}$.

The probability of rain tomorrow is 30%.

👍 EXAM TIP
Probabilities can be expressed as a fraction, decimal or percentage.

WORKED EXAMPLE

There are 7 red balls, 2 blue balls and 6 yellow balls in a bag.

A ball is chosen at random. Find the probability that the ball is:

a) blue [1 mark]
b) red [1 mark]
c) either red or blue. [1 mark]

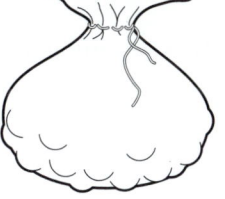

🔑 KEY SKILLS
You must be able to calculate simple probabilities.

There are 7 + 2 + 6 = 15 balls in the bag.

a) 2 out of the 15 balls are blue, so P(blue) = $\frac{2}{15}$

b) 7 out of the 15 balls are red, so P(red) = $\frac{7}{15}$

c) The events 'red' and 'blue' are mutually exclusive,
   so: P(red or blue) = $\frac{2}{15} + \frac{7}{15} = \frac{9}{15} = \frac{3}{5}$

👍 EXAM TIP
If you give your probability as a fraction, you can choose to simplify it, or leave it unsimplified.

8 Probability

WORKED EXAMPLE

The probability of the traffic lights being red is 0.3.

The probability of them being green is 0.5.

a) Is it more likely that the lights are red or green? [1 mark]

b) Find the probability that the lights are not red. [1 mark]

a) It is more likely that the lights are green since 0.5 is greater than 0.3.

b) P(not red) = 1 − P(red) = 1 − 0.3 = 0.7

RECAP

P(event not occurring) = 1 − P(event occurring)

Extended

KEY SKILLS

At extended level, you are expected to understand and use probability notation.

If P(red) is denoted by P(R), then P(not red) is denoted by P(R').

QUESTIONS

1. A fair six-sided cube with faces numbered 1 to 6 is rolled.

   Find the probability that it shows:
   a) an even number
   b) a prime number
   c) a number less than 5
   d) a number greater than 7.

2. A bag contains 7 blue marbles, 6 red marbles and 3 yellow marbles.

   One marble is picked at random from the bag.

   Find the probability that it is:
   a) yellow
   b) red
   c) blue or yellow
   d) green.

3. The letters of the word RWANDA are written on cards and placed into a hat.

   One card is drawn at random.

   Find the probability that the letter shown is:
   a) W
   b) A
   c) a consonant
   d) in the first half of the alphabet.

4. 16 tiles, numbered from 1 to 16, are placed in a bag.

   One tile is drawn at random and the number is recorded.

   Find the probability that it is:
   a) a factor of 16
   b) a multiple of 3
   c) a prime number
   d) less than 10
   e) greater than 17
   f) either a factor of 12 or a factor of 15.

5. A set of cards contains 45 cards.

All the cards are either red or black.

The probability of choosing a red card from the set is 0.6.

   a) What is the probability of choosing a black card?
   b) Which colour card is more likely to be chosen?
   c) How many red cards are there in the set?

6 A bag contains 52 counters, which are either red, green, blue or yellow. There are 13 counters of each colour in the bag. Each set of coloured counters is numbered from 1 to 13.

One counter is drawn at random.

Work out the probability that it is:

   a) a red counter
   b) a counter numbered 13
   c) a counter with a number less than 6
   d) a counter numbered 11 or 12
   e) a yellow counter numbered 12
   f) not a green counter.

To **Raise your grade** now try questions 4, 5 and 11 on pages 201–202

8 Probability

8.2 Relative frequency

YOU NEED TO:
- Understand relative frequency as an estimate of probability.
- Calculate expected frequencies.

◀◀ RECAP
The relative frequency of an event occurring is defined as

$$\text{relative frequency} = \frac{\text{number of times the event occurs}}{\text{total number of trials}}$$

🔑 KEY SKILLS
You must be able to calculate relative frequency.

WORKED EXAMPLE
Scarlett has a bag that contains 100 pens.

She knows that the bag contains only red, orange and purple pens.

Scarlett takes 20 pens from the bag at random, replacing the chosen pen each time.

Overall, she selects 8 pens that are red, 5 that are orange and 7 that are purple.

a) Write down the relative frequency of choosing a red pen. [1 mark]

b) Write down the relative frequency of **not** choosing an orange pen. [1 mark]

c) Write down an estimate for the number of pens of each colour in the bag. [2 marks]

She repeats the experiment but takes out 50 pens, 15 of which are red.

d) Explain how this would affect your answer to part **b)**. [1 mark]

a) Relative frequency $= \frac{8}{20} = \frac{2}{5}$

b) Relative frequency $= \frac{8+7}{20} = \frac{15}{20} = \frac{3}{4}$

c) Red: $\frac{2}{5} \times 100 = 40$ Orange: $\frac{5}{20} \times 100 = 25$ Blue: $\frac{7}{20} \times 100 = 35$

d) The estimate for red would be lower.

👍 EXAM TIP
Remember that 'relative frequency' is a probability that you obtain from data.

The more data you have, the more reliable the relative frequency is.

❓ QUESTIONS

1. Jenny rolls a cube with faces numbered from 1 to 6 18 times.
   She gets a 6 twice.

   a) If the number cube is fair, how many times would she have expected to get a 6?

   She rolls the number cube another 32 times and gets a 6 three more times.

   b) Based on this data, what is the probability of rolling a 6 with this number cube?

2. Kieran takes 12 sweets at random from a bag. Four of them are red.
   a) Based on this data, what is the probability that a sweet taken from the bag is red?
   b) If there are 60 sweets in the bag, how many would you expect to be red?

3. Derren wanted to test whether a particular coin he owned was a fair coin.
   a) If the coin was fair, and he flipped the coin 40 times, how many times should he expect it to land on heads?
   b) Derren flipped the coin 40 times and it landed on heads 17 times.

   He said, 'It isn't a fair coin, because 17 is less than the expected number of times.'

   How might you respond to this statement?

4. Lucy wanted to know how many students in her school owned a pet.

   She asked 15 students at random if they owned a pet and 10 said yes.
   a) Based on this data, what is the probability that a randomly selected student owns a pet?
   b) If there are 900 students in her school, how many of them should Lucy expect to own a pet?

   Lucy asks a different 15 students if they own a pet and this time only 8 say yes.
   c) Based on all the data combined, how many students would she now expect to own a pet?

To **Raise your grade** now try questions 2, 6 and 8 on pages 201–202

8 Probability

8.3 Combined events

> **YOU NEED TO:**
> - Calculate the probability of combined events using sample space diagrams, Venn diagrams and tree diagrams, as appropriate.

WORKED EXAMPLE

A red cube and a blue cube, both with faces numbered from 1 to 6, are rolled at the same time.

The sample space diagram displays all the possible outcomes.

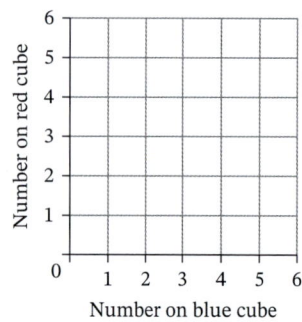

Find the probability of obtaining:

a) a total of 9 [1 mark]

b) a total of 3 [1 mark]

c) two sixes. [1 mark]

> 🔑 **KEY SKILLS**
> You must be able to calculate probabilities of combined events using a sample space diagram.

> ⏪ **RECAP**
> Sample space diagrams display all the possible outcomes of an event in a grid.

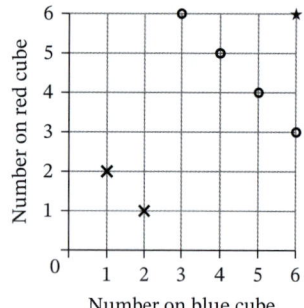

There are 36 possible outcomes, shown where the lines cross.

a) There are four ways of obtaining a total of 9; therefore,

$$P(\text{total of } 9) = \frac{4}{36} = \frac{1}{9}$$

b) There are two ways of obtaining a 3; therefore,

$$P(\text{total of } 3) = \frac{2}{36} = \frac{1}{18}$$

c) There is only one way of obtaining two sixes; therefore,

$$P(\text{two sixes}) = \frac{1}{36}$$

194

8.3 Combined events

◀◀ RECAP

Tree diagrams are a clear way of representing the possible outcomes of combined events.

On tree diagrams:

- as you move across, multiply probabilities
- as you move down, add probabilities.

🗝 KEY SKILLS

You must be able to calculate probabilities of combined events using a tree diagram.

WORKED EXAMPLE

A bag contains 10 balls, seven of which are red and the rest green.

A boy randomly takes out a ball from the bag, notes its colour and puts it back.

He then takes a second ball from the bag, noting its colour.

a) Draw a tree diagram to represent this information. **[2 marks]**
b) Find the probability that he chooses two red balls. **[2 marks]**
c) Find the probability that he chooses two balls of different colours. **[2 marks]**

👍 EXAM TIP

Remember to put the probabilities on the branches and the event labels at the end of the branches. The only exceptions to this rule are the final combined probabilities at the ends of the diagram.

a)

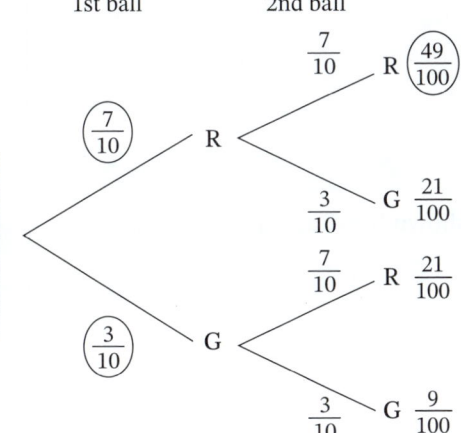

Notice that the probabilities on the branches that start from the same point always add up to 1.

b) $P(RR) = \dfrac{7}{10} \times \dfrac{7}{10} = \dfrac{49}{100}$

c) $P(RG) = \dfrac{7}{10} \times \dfrac{3}{10} = \dfrac{21}{100}$ and $P(GR) = \dfrac{3}{10} \times \dfrac{7}{10} = \dfrac{21}{100}$

The probability of choosing two balls of different colours is

$\dfrac{21}{100} + \dfrac{21}{100} = \dfrac{42}{100} = \dfrac{21}{50}$

⊙ WATCH OUT!

Questions often involve choosing two or more balls from a bag. In these questions it is important to establish whether or not the first ball has been replaced before the next one has been chosen. Read the question carefully. In this worked example, the balls are replaced.

◀◀ RECAP

Venn diagrams are sometimes useful for solving probability problems.

- The sets represent the outcomes of an event.
- The diagram can contain either probabilities or frequencies.

🗝 KEY SKILLS

You must be able to calculate probabilities of combined events using a Venn diagram.

8 Probability

WORKED EXAMPLE

In a class of 33 students, 20 like chess, 12 like draughts and 5 like neither.

a) Represent this information on a Venn diagram. [2 marks]
b) Find the probability that a randomly selected student:
   i) likes chess but not draughts [1 mark]
   ii) likes draughts but not chess. [1 mark]

a) The two circles can overlap. One circle has a total of 20 (students who like chess) and one has a total of 12 (students who like draughts).

The combined total has to be 33 − 5 = 28.

If the circles did not overlap, their combined total would be 32 (20 + 12), so the overlap must have 32 − 28 = 4 students in it.

There are 20 altogether who like chess, so 16 must go in the left circle.

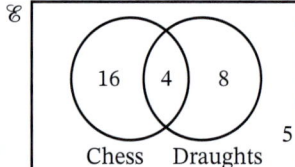

There are 12 altogether who like draughts, so 8 must go in the right circle.

b) i) P(chess but not draughts) = $\frac{16}{33}$

   ii) P(draughts but not chess) = $\frac{8}{33}$

? QUESTIONS

1. Two identical spinners, like the one shown here, are spun.

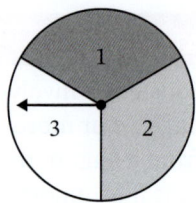

The area occupied by each number on the spinners is equal.

a) How many possible outcomes are there?
b) What is the probability of getting a 2 on both spinners?
c) What is the probability of getting an odd number on either spinner?
d) What is the probability of getting a total of 4?
e) What is the probability of getting a total of 1?

2. A bag contains 6 blue counters and 5 green counters.

A counter is drawn from the bag, its colour is noted and then it is replaced.

A second counter is then drawn.

a) Draw a tree diagram showing the possible outcomes.
b) What is the probability of drawing two blue counters?
c) What is the probability of drawing one counter of each colour?

8.3 Combined events

3. A road has two sets of traffic lights.

   The probability of being stopped at each set is 0.3.

   a) Draw a tree diagram showing the possible outcomes.
   b) What is the probability of getting stopped at both sets?
   c) What is the probability of getting stopped at neither set?

4. In a year of 100 students, 70 like Maths, 50 like French and 20 like neither.

   a) Draw a Venn diagram to display this information.
   b) Use your diagram to work out the probability that a randomly selected student likes both Maths and French.

5. A bag contains 4 green balls, 3 red balls and 2 yellow balls.

   A ball is chosen at random, its colour is noted and then it is replaced.

   A second ball is then chosen.

   a) Draw a tree diagram showing the possible outcomes.
   b) What is the probability that both balls are green?
   c) What is the probability that the balls are the same colour?

6. The probability that Sheila is late to school is 0.2.

   a) Draw a tree diagram showing all of the possibilities for Sheila's arrivals on three consecutive days.
   b) What is the probability that Sheila is late on all three days?
   c) What is the probability that she is late just once?

7. A bag contains 3 red balls, 4 green balls and 5 yellow balls.

   One ball is chosen at random from the bag, its colour is noted, and then it is replaced.

   A second ball is then chosen from the bag.

   Find the probability that:

   a) both balls are red
   b) both balls are green
   c) one ball is red and the other is yellow
   d) neither ball is yellow.

8. The letters from the word FANTASTIC are written on individual pieces of paper and placed in a basket.

   One letter is chosen at random and then replaced.

   A second letter is then chosen.

   Find the probability that:

   a) the letter A is obtained twice
   b) the letter N is obtained twice
   c) the letter F is selected first, followed by a vowel
   d) a consonant and a vowel are selected.

To **Raise your grade** now try questions 1, 3, 7, 9 and 10 on pages 201–202

8 Probability

8.4 Conditional probability

YOU NEED TO:

- **E** Calculate conditional probability using Venn diagrams, tree diagrams and tables.

Extended

WORKED EXAMPLE

Out of 30 students in a class, 20 like pop music, 14 like rock music and 4 like neither.

a) Illustrate this information using a Venn diagram. [2 marks]
b) Find the probability that a randomly selected student:
   i) likes pop music and rock music [2 marks]
   ii) likes pop music, given that they do not like rock music [2 marks]
   iii) likes rock music, given that they do not like pop music. [2 marks]

a)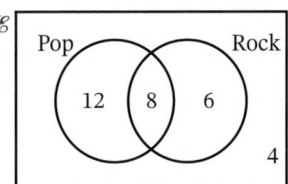

b) i) $P(\text{pop and rock}) = \dfrac{8}{30} = \dfrac{4}{15}$

   ii) There are 16 people who don't like rock music, and 12 of those like pop music; therefore, the probability that a student likes pop music, given that they do not like rock music is

   $P(\text{pop} \mid \text{rock}') = \dfrac{12}{16} = \dfrac{3}{4}$

   iii) There are 10 people who don't like pop music, and 6 of those like rock music; therefore, the probability that a student likes rock music, given that they do not like pop music is

   $P(\text{rock} \mid \text{pop}') = \dfrac{6}{10} = \dfrac{3}{5}$

KEY SKILLS

You must be able to calculate a conditional probability using a Venn diagram.

EXAM TIP

Conditional probability questions often contain the word 'given'.

EXAM TIP

The notation for the probability of 'A given B' is $P(A \mid B)$.

8.4 Conditional probability

WORKED EXAMPLE

There are 3 red balls and 2 green balls in a bag. One ball is chosen at random, its colour is noted and it is not replaced.

A second ball is then chosen and its colour is also noted.

a) Draw a tree diagram to illustrate this situation. [2 marks]

b) Find the probability that:
   i) both balls are red [2 marks]
   ii) both balls are the same colour [2 marks]
   iii) at least one of the balls is green [2 marks]
   iv) the first ball is red, given that the second ball is green. [2 marks]

🔑 KEY SKILLS

You must be able to calculate a conditional probability using a tree diagram.

a)

When the 5 balls are in the bag, the probability of choosing a red is $\frac{3}{5}$, since 3 of the 5 balls are red.

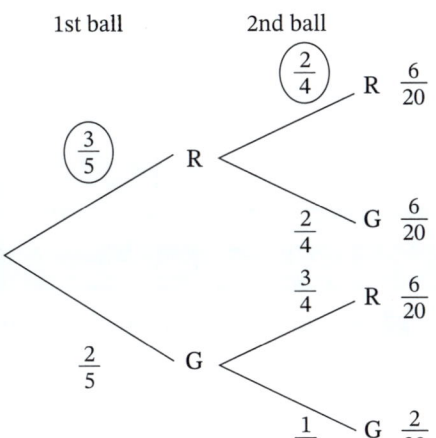

If a red ball is chosen first, then there are now 4 balls in the bag and only 2 of them are red. The probability of choosing a red is $\frac{2}{4}$.

👍 EXAM TIP

The fact that the first ball is not replaced means that the probability of the second ball being a particular colour is conditional and depends on the colour of the first ball.

b) The probabilities are:

i) $P(RR) = \frac{3}{5} \times \frac{2}{4} = \frac{6}{20} = \frac{3}{10}$

ii) $P(\text{same colour}) = P(RR \text{ or } GG)$

$P(\text{same colour}) = \frac{6}{20} + \frac{2}{20} = \frac{8}{20} = \frac{2}{5}$

iii) When two balls are chosen, either 'At least one of the balls is green' or both of them are red.

So the probabilities of these two events add up to 1.

$P(\text{at least one G}) + P(RR) = 1$

So $P(\text{at least one G}) = 1 - P(RR)$

Therefore $P(\text{at least one G}) = 1 - \frac{3}{10} = \frac{7}{10}$

iv) $P(\text{first ball red} \mid \text{second ball green})$

To calculate this, divide the probability of 'the first ball is red, then the second ball is green' by the probability that the second ball is green:

$P(\text{first ball red} \mid \text{second ball green}) = \dfrac{\frac{6}{20}}{\frac{6}{20} + \frac{2}{20}} = \frac{6}{8} = \frac{3}{4}$

👍 EXAM TIP

Part b) iii) shows an example of an important short cut.

You do not need to think of all the cases when there is at least one green ball.

✏️ APPLY

Think about how the method of solving part d) shown in the above worked example relates to the Venn diagram method.

199

8 Probability

WORKED EXAMPLE

Here is a table showing the meal choices made by 60 people at a wedding.

	Ice-cream	Cake	Brownie	Total
Chicken	9	16	12	37
Fish	6	12	5	23
Total	15	28	17	60

Find the probability that a randomly selected person:

a) chose a brownie, given that they chose chicken [2 marks]

b) chose chicken, given that they chose ice-cream. [2 marks]

a) The number of people who chose chicken was 37, and 12 of those people chose a brownie, so $P(\text{brownie}|\text{chicken}) = \dfrac{12}{37}$

b) The number of people who chose ice-cream was 15, and 9 of those people chose chicken, so $P(\text{chicken}|\text{ice-cream}) = \dfrac{9}{15} = \dfrac{3}{5}$

🔑 KEY SKILLS

You must be able to calculate a conditional probability using a table.

❓ QUESTIONS

1. A boy is late 60% of the time when it is raining and 30% of the time when it is dry. It rains on 25% of days.
   a) Draw a tree diagram to represent the above information.
   b) Find the probability that:
      i) it is raining and he is late
      ii) he is late
      iii) he is on time given that it is dry.

2. On an athletics day, 150 athletes take part. 60 are in the 100 metres race, 50 are in the 200 metres race and 80 are in neither.
   a) Draw a Venn diagram showing this information.
   b) Find the probability that a randomly selected athlete will be:
      i) taking part in both races
      ii) taking part in the 100 m race, given that they are taking part in the 200 m race
      iii) taking part in the 200 m race, given that they are taking part in the 100 m race.

3. At a café, customers can choose between tea and coffee.
   Both drinks come in three different sizes.
   The café sells 50 drinks as displayed in the following table.

	Small	Medium	Large	Total
Tea	12	14	9	35
Coffee	5	3	7	15
Total	17	17	16	50

Find the probability that a randomly selected customer will have chosen:

a) a small drink, given that they chose tea

b) coffee, given that they chose a large drink

c) a medium drink, given that they did not choose coffee.

To **Raise your grade** now try questions 12, 13 and 14 on page 203

↑ Raise your grade

1. Dipika rolls two cubes with faces numbered from 1 to 6.
   What is the probability that both number cubes land on a six? [2 marks]

2. The spinner shown is spun 400 times.

   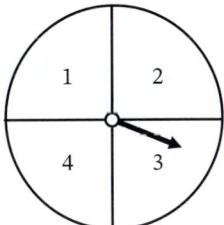

   How many times would you expect the spinner to land on:
   a) the sector numbered 1 [1 mark]
   b) a sector with an even number [1 mark]
   c) a sector with a number of more than 3? [1 mark]

3. At a sports club, 30 members are asked if they like hockey (H), rugby (R) or both.
   Some of the results are shown in the Venn diagram.

   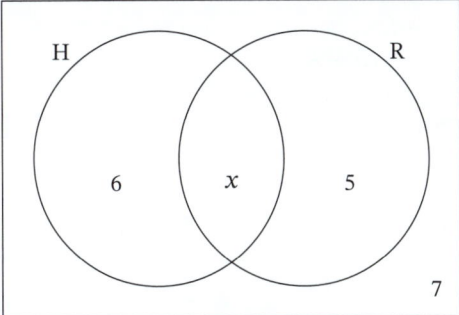

   a) Find the value of x. [2 marks]
   b) Two members of the club are chosen at random.
      Find the probability that both members like either hockey or rugby or both. [2 marks]

4. A spinner is equally likely to land on the numbers 1, 2, 3, 4, 5 or 6.
   Find the probability that the spinner lands on:

   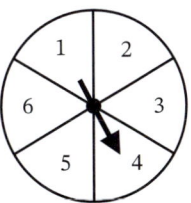

   a) an even number [1 mark]
   b) a number greater than 4 [1 mark]
   c) an even number or a number greater than 4. [1 mark]

5. The probability of a girl wearing glasses is 0.3 and the probability of a girl having blonde hair is 0.4.
   Joshua thinks that the probability of a girl wearing glasses or having blonde hair is 0.7.
   Is he correct? Explain your answer. [2 marks]

8 Probability

6. Mr Choudry drives to work each morning.

    The probability that he parks his car at the front of the building is 0.4.

    The probability that he parks his car at the side of the building is 0.15.

    The rest of the time he parks at the back of the building.

    a) What is the probability that Mr Choudry will park at the back of the building on any particular morning? [1 mark]

    b) What is the probability that he will park either at the back or at the side of the building on any particular morning? [1 mark]

    In the next year, he works for 220 days.

    c) On approximately how many days will Mr Choudry park either at the back or at the side of the building? [2 marks]

7. Alexander shoots three arrows at a target.

    With each arrow, the probability that he hits the target is 0.3.

    Whether or not he hits the target with any given arrow is independent of what happened to the previous arrows.

    Find the probability that he hits the target with the first two arrows but misses with the third. [2 marks]

8. A boy picks out a marble from a bag of 20 coloured marbles.

    He records its colour and then puts it back.

    He does this fifty times.

    Ten of the marbles he takes out are red.

    Estimate how many of the 20 marbles in the bag are red. [2 marks]

9. A box contains 12 counters of which two are blue, three are red and the rest are purple.

    Two counters are chosen at random, the first being replaced before the second is chosen.

    Find the probability that

    a) the first counter is purple and the second is blue [1 mark]

    b) both counters are red [2 marks]

    c) one of the counters is blue and the other is red. [2 marks]

10. A coin is biased so that heads appears twice as often as tails.

    a) If p is the probability of getting tails, write down the probability of getting heads in terms of p. [1 mark]

    b) Hence, show that $p = \frac{1}{3}$. [2 marks]

    c) The coin is tossed twice. Find the probability of getting:

       i) two heads

       ii) a head first and then a tail

       iii) two tails. [4 marks]

11. A circular spinner has four sections, numbered 1, 2, 3 and 4.

    The area for 2 is twice the area for 1, the area for 3 is three times the area for 1 and the area for 4 is four times the area for 1.

    If p is the probability of getting a 1 then:

    a) write down, in terms of p, the probabilities of getting 2, 3 and 4 [2 marks]

    b) find p [2 marks]

    c) find the probability of getting an even number. [2 marks]

Extended

12. Mr Danyata travels to school by car.

    The probability that he is delayed in traffic on a particular day is 0.4.

    If he is delayed in traffic, the probability that he is late to school is 0.7.

    If he is not delayed in traffic, the probability that he is late is 0.05.

    Find the probability that Mr Danyata is late to school. **[3 marks]**

13. Five yellow balls and three red balls are placed in a bag and two are removed, one at a time, without replacement.

    a) Draw a tree diagram to represent this information. **[3 marks]**

    b) Find the probability that:
        i) both balls are red
        ii) both balls are the same colour
        iii) the second ball is yellow given that the first is red. **[6 marks]**

14. Fifty people attending a conference were offered a choice of tea, coffee or hot chocolate, along with biscuits or a piece of cake.

    Their choices are displayed in this table.

	Tea	Coffee	Chocolate	Total
Biscuits	12	6	8	26
Cake	10	9	5	24
Total	22	15	13	50

    Work out the probability that a randomly selected member of the group:

    a) chose cake **[1 mark]**

    b) chose coffee **[1 mark]**

    c) chose biscuits, given that they chose tea **[1 mark]**

    d) chose tea, given that they chose biscuits. **[1 mark]**

9 Statistics

Your revision checklist

Tick these circles to build a record of your revision.

Core/ **E** Extended syllabus

		☹ 😐 🙂
9.1	Classify and tabulate statistical data.	○ ○ ○
9.2	Read, interpret and draw inferences from tables and statistical diagrams.	○ ○ ○
	Compare data using tables, graphs and statistical measures.	○ ○ ○
	Appreciate restrictions on drawing conclusions from given data.	○ ○ ○
9.3	Calculate the mean, median, mode and range for individual data and distinguish between the purposes for which they are used.	○ ○ ○
	E Calculate the quartiles and interquartile range for data.	○ ○ ○
	E Calculate an estimate of the mean for grouped discrete or grouped continuous data.	○ ○ ○
	E Identify the modal class from a grouped frequency distribution.	○ ○ ○
9.4	Construct and interpret bar charts, pie charts, pictograms, stem-and-leaf diagrams and simple frequency distributions.	○ ○ ○
9.5	Construct and interpret scatter diagrams.	○ ○ ○
	Understand what is meant by positive, negative and zero correlation.	○ ○ ○
	Draw and interpret straight lines of best fit by eye.	○ ○ ○
9.6	**E** Construct and interpret cumulative frequency tables and diagrams.	○ ○ ○
	E Estimate and interpret the median, percentiles, quartiles and interquartile range from cumulative frequency diagrams.	○ ○ ○
9.7	**E** Draw and interpret histograms.	○ ○ ○
	E Calculate with frequency density.	○ ○ ○

9.1, 9.2, 9.4, 9.7 Displaying data

YOU NEED TO:

- Classify and tabulate statistical data.
- Read, interpret and draw inferences from tables and statistical diagrams.
- Compare data using tables, graphs and statistical measures.
- Appreciate restrictions on drawing conclusions from given data.
- Construct and interpret bar charts, pie charts, pictograms, stem-and-leaf diagrams and simple frequency distributions.
- Ⓔ Draw and interpret histograms.
- Ⓔ Calculate with frequency density.

WORKED EXAMPLE

The homework scores of 20 students were as follows:

20	20	15	20	17
16	18	18	20	15
20	18	19	17	18
20	19	18	20	18

Represent this information using a bar chart. **[3 marks]**

Start by putting the scores into a frequency table:

Score	Frequency
15	2
16	1
17	2
18	6
19	2
20	7

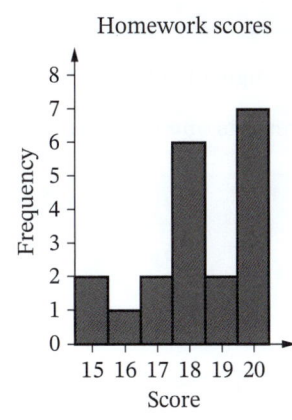

Then use the table to draw the bar chart.

🔙 RECAP

In a pie chart, you use sectors to represent the data.

An angle of x in the pie chart represents $\dfrac{x}{360}$ of the total.

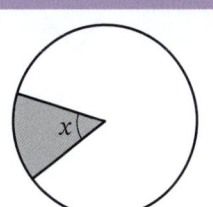

🔑 KEY SKILLS

You need to be able to tabulate both discrete and continuous data, read and interpret various statistical diagrams, and compare sets of data, drawing conclusions while appreciating the restrictions involved.

🔑 KEY SKILLS

You must be able to construct and interpret bar charts.

🔙 RECAP

In a bar chart, the frequency is represented by the height of the bar.

👍 EXAM TIP

Frequency always goes on the vertical axis.

🔑 KEY SKILLS

You must be able to construct and interpret pie charts.

205

9 Statistics

WORKED EXAMPLE

The pie chart shows the answers that 120 students gave to a question on a multiple choice paper.

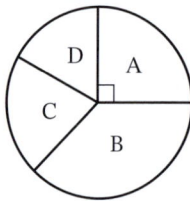

How many students put A as the answer? **[2 marks]**

The right angle in the sector representing A shows that $\frac{90}{360}$ or one-quarter of the students chose A as their answer.

$\frac{1}{4}$ of $120 = \frac{1}{4} \times 120 = 30$

30 students gave the answer A.

WORKED EXAMPLE

James asked 36 students in his class where they had been on holiday. He recorded the results in this table:

Country	UK	France	Spain	Other
Number of students	9	6	11	10

Represent this information in a pie chart. **[3 marks]**

At the centre of the pie there is an angle of 360°.

There are 36 students, so 10° represents one student.

Country	UK	France	Spain	Other
Number of students	9	6	11	10
Angle of sector	90°	60°	110°	100°

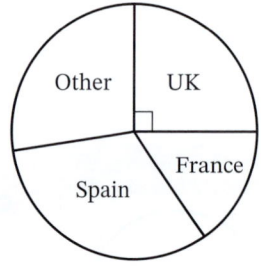

9.1, 9.2, 9.4, 9.7 Displaying data

WORKED EXAMPLE

The number of students in each year at a school are as follows:

Year	Frequency
7	80
8	75
9	70
10	70
11	100

Represent this information in a pictogram. [4 marks]

RECAP

In a pictogram, you use symbols or pictures to represent data.

KEY SKILLS

You must be able to draw and interpret pictograms.

The symbol 👤 represents 10 students.

Year	Frequency
7	👤 👤 👤 👤 👤 👤 👤 👤
8	👤 👤 👤 👤 👤 👤 👤 ½
9	👤 👤 👤 👤 👤 👤 👤
10	👤 👤 👤 👤 👤 👤 👤
11	👤 👤 👤 👤 👤 👤 👤 👤 👤 👤

EXAM TIP

Note that half a person represents 5 students.

WORKED EXAMPLE

A class of 25 students all took a history test that was marked out of 50. Here are their marks.

22	18	8	42	25	35	47	33	28	9
25	13	46	50	31	44	26	10	49	17
39	21	30	27	34					

Display this data in a stem-and-leaf diagram. [4 marks]

Use the stem part of the diagram to represent the tens digit, and the leaf part to represent the ones digit.

Stem	Leaf						Key 1|3 means 13
0	8	9					
1	0	3	7	8			
2	1	2	5	5	6	7	8
3	0	1	3	4	5	9	
4	2	4	6	7	9		
5	0						

KEY SKILLS

You must be able to construct and interpret stem-and-leaf diagrams.

RECAP

A stem-and-leaf diagram is a way of displaying data in groups.

EXAM TIP

This kind of stem-and-leaf diagram is called an 'ordered stem-and-leaf diagram' because the numbers in the leaf section are in ascending order. It is sometimes a good idea to make an unordered diagram first, particularly if the data set is large.

9 Statistics

WORKED EXAMPLE

Two teams, of nine runners each, competed in a 1500 metre race. Their times were as follows.

Team 1: 7 m 11 s, 7 m 15 s, 7 m 30 s, 7 m 43 s, 8 m 22 s, 8 m 37 s, 8 m 38 s, 9 m 1 s, 9 m 10 s

Team 2: 6 m 20 s, 6 m 48 s, 6 m 51 s, 7 m 9 s, 7 m 20 s, 8 m 12 s, 8 m 22 s, 8 m 35 s, 8 m 49 s

Display this data in a back-to-back stem-and-leaf diagram. **[4 marks]**

Construct the diagram so that Team 1 goes to the left and Team 2 goes to the right.

	Team 1				Team 2			
Key (Team 1)				6	20	48	51	Key (Team 2)
43 | 7 means	43	30	15	11	7	09	20	7 | 20 means
7 min 43 sec		38	37	22	8	12	22	35	49	7 min 20 sec
			10	01	9			

> ◀◀ **RECAP**
> Sometimes stem-and-leaf diagrams can be double sided, if you are trying to compare two sets of data. These are called back-to-back stem-and-leaf diagrams.

> 👁 **WATCH OUT!**
> You must always remember to include a key with your diagram.

> 👍 **EXAM TIP**
> Notice how the key is reversed on the left-hand side of the diagram.

◀◀ RECAP

A histogram is a kind of bar chart, with the following general features:

- They are used for continuous data.
- The bars often have different widths.
- The vertical axis is often labelled 'frequency density'.
- There are no gaps between the bars (unless there is a bar with zero height).
- The horizontal axis is labelled, not the bars.
- The frequency is determined by the area of the bar.

> 🔑 **KEY SKILLS**
> You must be able to construct and interpret histograms with equal class widths.

WORKED EXAMPLE

The table shows the distribution of the ages of 100 people attending a music festival.

Age	Frequency
$0 \leq x < 20$	22
$20 \leq x < 40$	45
$40 \leq x < 60$	19
$60 \leq x < 80$	14

Draw a histogram to display the data. **[4 marks]**

> 👍 **EXAM TIP**
> A histogram with equal class intervals is not much different to a regular bar chart.

9.1, 9.2, 9.4, 9.7 Displaying data

The ages of people at a music festival

[Histogram showing frequency of ages 0-80]

EXAM TIP

Notice that the bars are not labelled. Only the axes are labelled.

The bars have equal widths because the class intervals are all the same size.

The vertical axis is labelled 'Frequency' because the bars have equal widths.

Extended

WORKED EXAMPLE

The lengths of time 120 cars were parked in a short stay car park are recorded in the following table.

Time (min)	Frequency
$0 \leq x < 5$	10
$5 \leq x < 15$	20
$15 \leq x < 30$	45
$30 \leq x < 60$	30
$60 \leq x < 90$	15

Display this data in a histogram. **[6 marks]**

First, work out the frequency density for each class interval.

Time (min)	Frequency	Frequency density
$0 \leq x < 5$	10	2
$5 \leq x < 15$	20	2
$15 \leq x < 30$	45	3
$30 \leq x < 60$	30	1
$60 \leq x < 90$	15	0.5

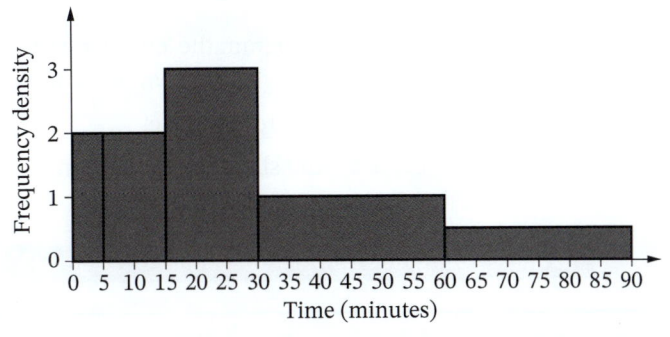

Minutes spent in a car park

KEY SKILLS

You must be able to construct and interpret histograms with unequal class widths.

RECAP

Most histogram questions will be about histograms with unequal class intervals.

EXAM TIP

Frequency density = frequency ÷ class width

The frequencies are represented by the area of each bar.

209

9 Statistics

QUESTIONS

1. Draw a bar chart for this data set.

Colour of car	Frequency
Red	1
Blue	12
Silver	8
Black	2

2. The pie chart shows the nationalities of a group of 150 delegates at a conference.

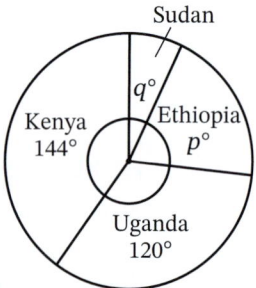

Given that 30 delegates came from Ethiopia, find:
   a) the value of p and the value of q
   b) how many delegates came from each of Kenya, Uganda and Sudan.

3. 11 students are asked to close their eyes for what they think is 60 seconds.

   Their actual times in seconds are:

   70, 64, 49, 52, 61, 53, 72, 57, 53, 67, 48

   Draw a stem-and-leaf diagram to display this data.

Extended

4. The prices of the goods (to the nearest $) sold by an electrical shop on a certain day are summarised below.

Price of goods ($)	Frequency	Frequency density
$20 \leq x < 40$	20	1
$40 \leq x < 50$	37	
$50 \leq x < 60$		6.2
$60 \leq x < 70$	51	
$70 \leq x < 90$	30	
$90 \leq x < 130$		0.2

   a) Copy and complete the table.
   b) Draw a histogram to illustrate these data using 1 cm per $10 on the horizontal axis (from 20 to 130) and 2 cm per unit on the vertical axis (from 0 to 7).

5. A student drew a histogram for the data shown in the table below. They added a column to the table to show the width of each bar.

Mass (kg)	Frequency	Frequency density (height of bar)	Width of bar
10–	8		
20–	16		
50–	38		
60–	40	4 cm	1 cm
70–	28		
75–	30		
80–	21		
85–	11		
90–120	6		

   On the student's histogram the '60–' bar was 1 cm wide and 4 cm high as shown in the table.
   a) Copy and complete the table.
   b) Hence, draw the student's histogram using the scale that they used.

To **Raise your grade** now try question 9 on page 223

9.3 Averages and range

> **YOU NEED TO:**
> - Calculate the mean, median, mode and range for individual data and distinguish between the purposes for which they are used.
> - **E** Calculate the quartiles and interquartile range for data.
> - **E** Calculate an estimate of the mean for grouped discrete or grouped continuous data.
> - **E** Identify the modal class from a grouped frequency distribution.

> **◀◀ RECAP**
>
> The **mean** is 'the sum of all the data values' divided by 'the number of values', which can be written as $\bar{x} = \dfrac{\sum x}{n}$.
>
> The **median** is the middle value when the data is written in ascending (or descending) order. If there are n values, the median is the $\dfrac{n+1}{2}$th value.
>
> The **mode** is the value which appears most frequently.
>
> The **range** is the difference between the largest and smallest values.

> **🗝 KEY SKILLS**
>
> You must be able to calculate the mean, median, mode and range for individual and discrete data.

> **👍 EXAM TIP**
>
> Mean, median and mode are all averages, and are a way of representing a set of numbers using only one number. The range tells you how spread out the data is.

WORKED EXAMPLE

Here are some data:

14, 16, 10, 15, 13, 13, 19, 15, 13, 12, 14

Find:

a) the mean   [2 marks]

b) the median   [1 mark]

c) the mode   [1 mark]

d) the range.   [1 mark]

a) Mean $= \dfrac{14 + 16 + 10 + 15 + 13 + 13 + 19 + 15 + 13 + 12 + 14}{11}$

$= \dfrac{154}{11} = 14$

b) First, write the data in ascending order:

10, 12, 13, 13, 13, 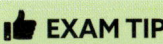, 14, 15, 15, 16, 19

Median = 14

c) Mode = 13

d) Range = 19 − 10 = 9

> **👍 EXAM TIP**
>
> There are 11 values, so the median is the $\dfrac{11+1}{2}$ = 6th value.

9 Statistics

Extended

> ### ◀◀ RECAP
>
> The **lower quartile** is the data value one-quarter of the way through the data set. It can be thought of as the median of the lower half of the data set.
>
> The **upper quartile** is the data value three-quarters of the way through the data set. It can be thought of as the median of the upper half of the data set.
>
> The **interquartile range** is the difference between the upper quartile and the lower quartile.

WORKED EXAMPLE

Find the interquartile range for the data set in the previous example. **[2 marks]**

The lower half of the data set is:

10, 12, ⑬, 13, 13

The middle number is 13, so the lower quartile is 13.

The upper half of the data set is:

14, 15, ⑮, 16, 19

The middle number is 15, so the upper quartile is 15.

The interquartile range = 15 − 13 = 2.

WORKED EXAMPLE

30 students are asked how many siblings they have. The results are shown in the table.

Number of siblings	0	1	2	3	4	5
Frequency	5	8	10	4	2	1

Calculate:

a) the mean number of siblings **[2 marks]**

b) the range of the number of siblings. **[1 mark]**

---

a) First calculate the sum of the number of siblings multiplied by the frequency:

$(0 \times 5) + (1 \times 8) + (2 \times 10) + (3 \times 4) + (4 \times 2) + (5 \times 1) = 53$

There are 30 students, so mean = 53 ÷ 30 = 1.77

b) Range = 5 − 0 = 5

> ### 🔑 KEY SKILLS
> You must be able to calculate the mean of a set of grouped data.

> ### ◀◀ RECAP
> You will sometimes need to calculate the mean from data in a table.

> ### 👁 WATCH OUT!
> Note that the mean of a set of whole numbers can be a decimal number. In fact, when using real data, it is a decimal most of the time.

> ### 👁 WATCH OUT!
> Make sure that you use the highest and lowest values of the data to calculate the range, rather than the highest and lowest frequencies.

9.3 Averages and range

WORKED EXAMPLE

The table shows the number of pages in a randomly selected set of 25 books.

Number of pages	$0 \leq x < 100$	$100 \leq x < 250$	$250 \leq x < 500$	$500 \leq x < 1000$
Frequency	5	10	8	2

a) Calculate an estimate of the mean number of pages in the books. **[2 marks]**

b) Why is your answer to part a) only an estimate? **[1 mark]**

c) Which is the modal class? **[1 mark]**

a) Because the data has been given in classes, you need to use the mid-point of each class as an estimate.

Calculate the sum of the number of pages multiplied by the frequency:

$(50 \times 5) + (175 \times 10) + (375 \times 8) + (750 \times 2) = 6500$

There are 25 books, so the mean number of pages = $6500 \div 25 = 260$.

b) Because you do not know how many pages any of the books have.

c) The modal class is $100 \leq x < 250$.

> **KEY SKILLS**
> You must be able to identify the modal class from a grouped frequency distribution.

> **WATCH OUT!**
> The modal class is not 10. That is the frequency.

QUESTIONS

 1. Here is a set of data:

7, 5, 2, 10, 10, 8, 11

Find:
a) the mean
b) the median
c) the mode
d) the range.

 2. Jane took five maths tests. Her scores were:

12, 25, 14, 22, 17

Find:
a) her mean score
b) her median score
c) the range of her scores.

Jane then took another test, and her mean score is now 19.

d) What did she score in the sixth test?

 3. The average height of 5 students in a class is 152 cm.

Another student whose height is 146 cm joins the class.

What is the new average height?

 4. The mean of 16 numbers is 17.5.

The mean of 12 of the numbers is 17.

What is the mean of the other 4 numbers?

9 Statistics

Extended

5. A cube with faces numbered 1 to 6 was rolled many times and its score (x) is shown in the table.

x	1	2	3	4	5	6
f	3	8	13	10	6	5

a) Find the mean of the frequency distribution.

Give your answer to 3 significant figures.

b) Find the median of the frequency distribution.

c) Write down the mode and the range of the frequency distribution.

6. The scores of some students in a test are given in the table.

Score (x)	11	12	13	14	15	16
Frequency (f)	2	5	7	6	3	1

Find:
a) the mean
b) the median
c) the mode
d) the range.

7. The table shows the masses, in grams, of 100 oranges.

Mass, x (g)	120–	130–	140–	150–	160–	170–180
Frequency	25	19	23	16	12	5

Find:
a) an estimate of the mean
b) the interval that contains the median
c) the modal class.

8. The masses of some elephants are recorded in the table below.

Mass, w (kg)	Frequency
$2000 \leq w < 2500$	7
$2500 \leq w < 3000$	10
$3000 \leq w < 3500$	15
$3500 \leq w < 4000$	21
$4000 \leq w < 4500$	13
$4500 \leq w < 5000$	7
$5000 \leq w < 5500$	4
$5500 \leq w < 6000$	3

a) Find an estimate for the mean weight.
b) Write down the median class.
c) Calculate an estimate for the range.

To **Raise your grade** now try questions 1, 2, 3, 4, 7 and 8 on pages 221–223

9.5 Scatter diagrams

> **YOU NEED TO:**
> - Construct and interpret scatter diagrams.
> - Understand what is meant by positive, negative and zero correlation.
> - Draw and interpret straight lines of best fit by eye.

◀◀ RECAP

You can use a scatter diagram to find out if there is any correlation (connection) between two sets of variables, such as shoe size and height, or sales of gloves and the outdoor temperature.

WORKED EXAMPLE

During the first week of the summer holidays, Jessie wrote down the temperature in her garden at noon, and also the number of hours she spent watching TV that day.

Temperature (°C)	28	27	20	22	18	14	25
Hours of TV	1	1.5	3	4	4.5	6	2

Draw a scatter diagram to illustrate this data. [4 marks]

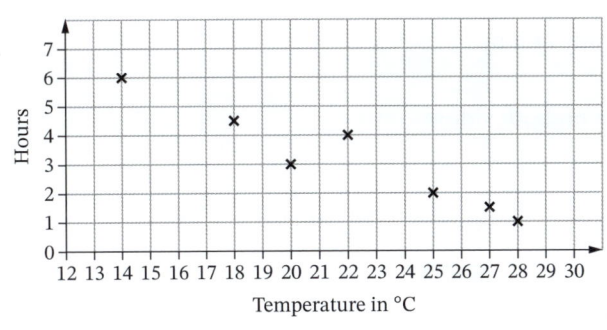

🔑 KEY SKILLS

You must be able to construct and interpret scatter diagrams.

👍 EXAM TIP

Because you suspect that the number of hours spent watching TV may depend on the temperature, you put temperature on the horizontal axis.

◀◀ RECAP

On a scatter diagram, 'correlation' means the way in which the two variables are related.

These points all lie close to a straight line with positive gradient.

There is said to be 'strong (linear) positive correlation'.

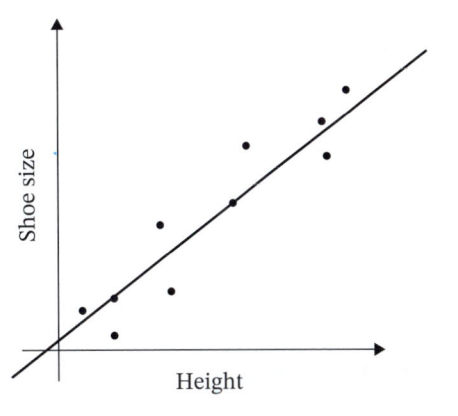

🔑 KEY SKILLS

You must be able to determine the nature of a correlation by examining a scatter diagram.

👍 EXAM TIP

Looking for a correlation is the main reason for drawing a scatter diagram.

9 Statistics

These points all lie close to a straight line with negative gradient.

There is said to be 'strong (linear) negative correlation'.

This is called the 'line of best fit'.

These points are spread out all over the graph.

There is said to be 'no (linear) correlation'.

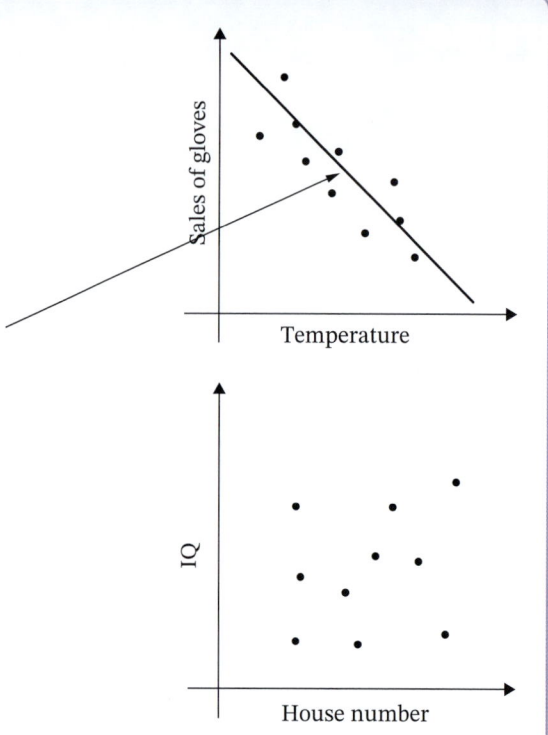

👍 EXAM TIP

Drawing the line of best fit 'by eye' just means 'without calculating the line's equation'. It means deciding on a sensible place for the line just by looking at the graph.

- Make sure your line is a straight line (always use a ruler!).
- Make sure there are no points you cannot get to from your line by moving vertically.
- Make sure there are roughly the same amount of points on both sides of the line.

WORKED EXAMPLE

The scatter diagram shows the number of hours a child spent online and the temperature at noon, over a seven-day period.

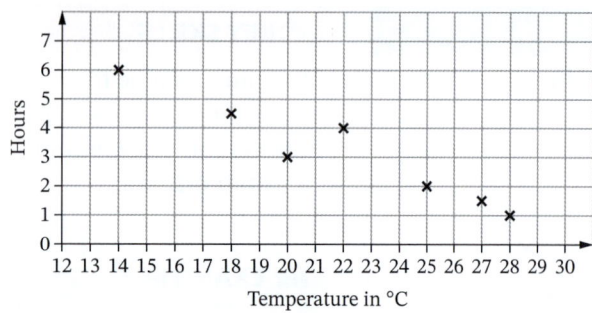

a) Draw a line of best fit by eye. [1 mark]
b) Describe the correlation. [1 mark]
c) Give an interpretation of the correlation. [1 mark]
d) If the temperature on one of the days had been 21°C, roughly how many hours would you expect the child to have spent online? [1 mark]

a)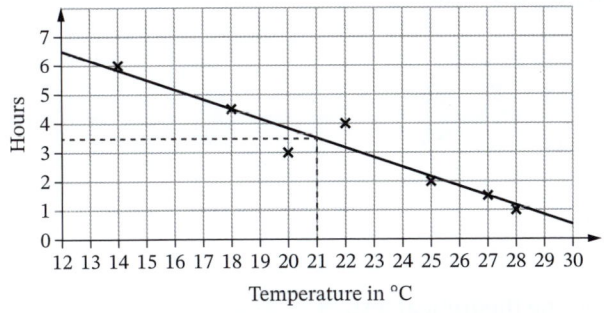

b) The scatter graph shows negative correlation.

c) The higher the temperature, the less time spent online.

d) 3.5 hours.

QUESTIONS

1. Describe the correlation shown in the scatter diagram.

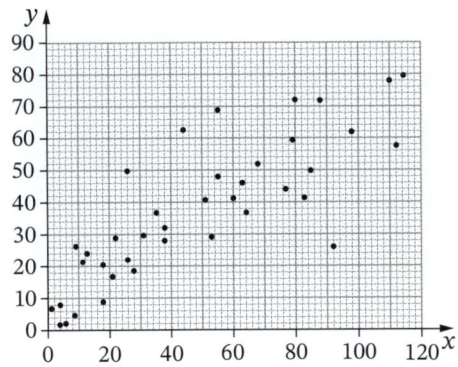

2. The table shows the speed of a motorcycle in miles per hour and its petrol consumption in miles per gallon.

   a) Draw a scatter diagram to show this information.
   b) Draw the line of best fit.

Speed (mph)	20	25	30	35	45	50	56	70
Petrol consumption (mpg)	80	76	70	64	54	50	47	40

   c) Estimate the fuel consumption when the motorcycle is travelling at 40 mph.
   d) State, with a reason, whether or not it is sensible to predict the fuel consumption when the motorcycle is travelling at 120 mph around a racetrack.

EXAM TIP

You can see the correlation easily by looking at the diagram. It does not need to be worked out.

EXAM TIP

Using the line of best fit to make predictions within the range of the given data is called 'interpolation' and is considered to be reliable, but making predictions outside the range of the given data is called 'extrapolation' and is considered to be unreliable.

WATCH OUT!

Remember that a correlation between two variables does not necessarily mean that one causes the other.

9 Statistics

3. The table shows the marks scored by 10 students in practical and theoretical biology tests.

Practical	45	36	14	22	25	31	44	38	27	36
Theoretical	39	33	18	21	27	29	48	43	27	32

a) Draw a scatter diagram to show this information.

b) Draw the line of best fit.

c) Another student took the practical test but missed the theoretical test.

If she scored 28 on the practical test, use your line of best fit to estimate her mark on the theoretical test.

d) State, with a reason, whether or not it is sensible to predict the practical test score of a student who scored 65 marks on the theoretical test.

To **Raise your grade** now try question 5 on page 222

9.6 Cumulative frequency

YOU NEED TO:

- **E** Construct and interpret cumulative frequency tables and diagrams.
- **E** Estimate and interpret the median, percentiles, quartiles and interquartile range from cumulative frequency diagrams.

Extended

WORKED EXAMPLE

The table shows the masses, in grams, of 100 apples.

Mass, x (g)	120–	130–	140–	150–	160–	170–180
Frequency	25	19	23	16	12	5

a) Draw a frequency table for this data, with a column for cumulative frequency. **[2 marks]**

b) Draw a cumulative frequency diagram for this data. **[4 marks]**

c) Estimate:
   i) the median **[1 mark]**
   ii) the 80th percentile **[1 mark]**
   iii) the interquartile range. **[2 marks]**

a)

Mass, x (g)	Frequency	Cumulative frequency
$120 \leq x < 130$	25	**25**
$130 \leq x < 140$	19	$25 + 19 =$ **44**
$140 \leq x < 150$	23	$44 + 23 =$ **67**
$150 \leq x < 160$	16	$67 + 16 =$ **83**
$160 \leq x < 170$	12	$83 + 12 =$ **95**
$170 \leq x < 180$	5	$95 + 5 =$ **100**

b)

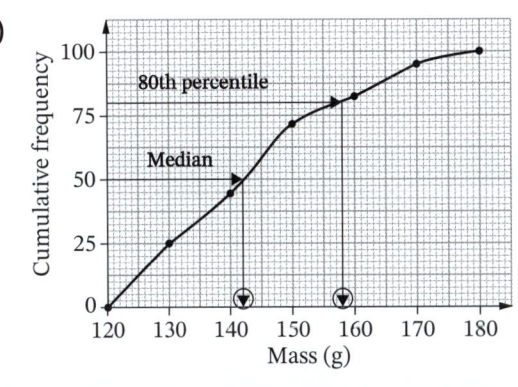

The curve will generally be smooth and must always have a positive gradient.

🔑 KEY SKILLS

You must be able to construct and interpret a cumulative frequency diagram.

◀◀ RECAP

A cumulative frequency diagram is an effective way to find estimates of the median and quartiles of grouped data.

👁 WATCH OUT!

Remember that each cumulative frequency is plotted at the upper end of its class interval.

👍 EXAM TIP

You should always join the points up with a curve if you can.

219

c) i) From the diagram, the median ≈ 142 (to 3 s.f.)

ii) From the diagram, the 80th percentile ≈ 158 (to 3 s.f.)

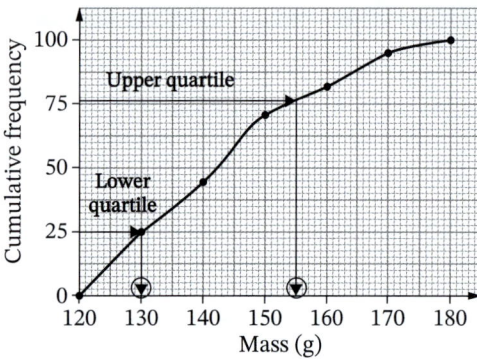

iii) From the diagram, the upper quartile ≈ 155 (to 3 s.f.)

The lower quartile ≈ 130 (to 3 s.f.)

The interquartile range ≈ 155 − 130 = 25.

QUESTIONS

1. The masses (in kg) of 100 animals are recorded in the table.

Mass (kg)	20–	25–	30–	35–	40–	45–50
Frequency	8	21	28	19	15	9

Write down the coordinates of the points through which the cumulative frequency curve should pass.

2. A biologist recorded the lengths of a group of 200 insects and displayed the results in the table.

Length (mm)	10–	11–	12–	13–	14–	15–	16–17
Frequency	17	38	53	49	21	12	10

a) Draw a cumulative frequency graph for these data using a scale of 2 cm per 1 mm on the horizontal axis and 1 cm per 10 units on the vertical axis. (The horizontal axis should go from 10 mm to 17 mm.)

b) Use the graph to find an estimate of the median.

c) Use the graph to find estimates of the lower and upper quartiles.

To **Raise your grade** now try question 6 on page 222

↑ Raise your grade

1. Here is a set of data:

   6, 8, 11, 14, 15, 15, 17

   Find:

   a) the mean [1 mark]

   b) the median [1 mark]

   c) the mode [1 mark]

   d) the range. [1 mark]

2. Twenty students were asked to estimate a time interval of 30 seconds.

   The stem-and-leaf diagram shows their actual times in seconds.

   Time in seconds

   ```
   1 | 9
   2 | 1 3 3 4 6 7 7 9
   3 | 0 1 1 2 2 2 4 5 6 6
   4 | 0
   ```

   Key 1 | 9 = 19 seconds

   Write down:

   a) the mode [1 mark]

   b) the median [1 mark]

   c) the range. [1 mark]

3. Here is a set of data:

   1.6, 1.2, 1.6, 1.3, 1.4, 1.5, 1.6, 1.7

   Find:

   a) the mean [1 mark]

   b) the median [1 mark]

   c) the mode [1 mark]

   d) the range. [1 mark]

4. The scatter diagram below shows the performance of a group of students in their autumn term and summer term examinations.

   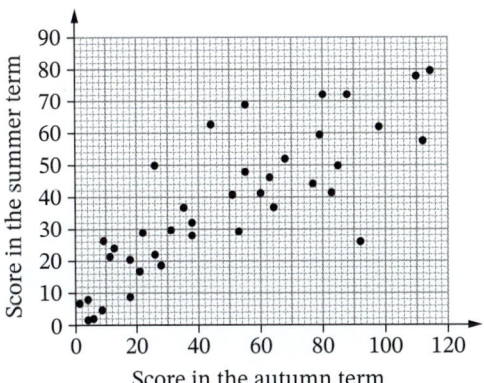

   a) Draw in a line of best fit by eye. [1 mark]

   b) What type of correlation does the diagram show? [1 mark]

   c) One student was absent for the summer term examination. If they scored 45 in the autumn term, what score do you predict they would get in the summer term? [2 marks]

9 Statistics

Extended

5. The marks scored by 49 students in a test are recorded in the table below.

Score, x	Frequency, f
12	1
13	7
14	8
15	5
16	6
17	8
18	12
19	2

Find:
- a) the mean score [2 marks]
- b) the median score [2 marks]
- c) the modal score. [1 mark]

6. The cumulative frequency curve shows the mass of African civets in a tropical rainforest.

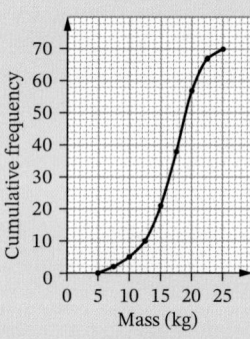

- a) Estimate the interquartile range of the masses. [3 marks]
- b) Estimate the number of civets with a mass greater than 21 kg. [2 marks]

7. The mass of 120 African bushpigs is shown in the table.

Mass, m (kg)	Frequency
$60 \leq m < 90$	5
$90 \leq m < 110$	24
$110 \leq m < 120$	51
$120 \leq m < 130$	31
$130 \leq m < 150$	9

Find an estimate for the mean mass of the bushpigs. [3 marks]

8. Giraffe heights at a zoo are recorded in the table below.

Height, h (m)	Frequency
$3.5 \leq h < 4.0$	12
$4.0 \leq h < 4.5$	18
$4.5 \leq h < 5.0$	15
$5.0 \leq h < 5.5$	8

Find:
a) an estimate for the mean height [3 marks]
b) the modal class. [1 mark]

9. At a school, a random sample of students was taken and each student recorded their intake of milk (in ml) during a given day.

Some of the results and part of the histogram are shown.

Milk intake (ml)	10–	30–	60–	100–	150–	200–	300–400
Number of students	4	9	32		25		
Frequency density							

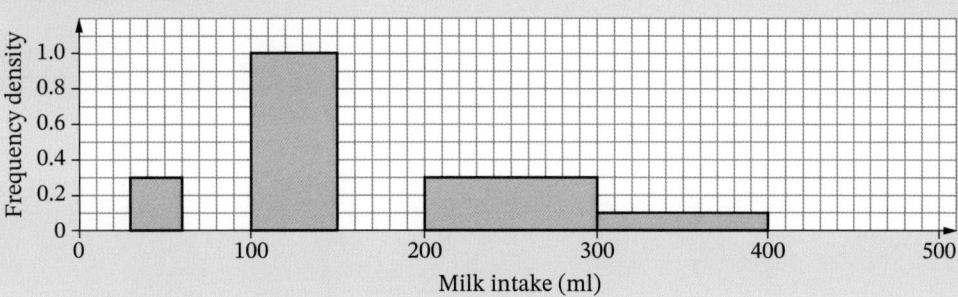

a) Calculate the frequency density for the '30–' category. [2 marks]
b) Copy the table, calculating the frequency density in all the cases where the number of students is stated. [2 marks]
c) Copy and complete the histogram to illustrate these data using 1 cm per 50 ml on the horizontal axis and 1 cm per 0.1 on the vertical axis. Use the table to complete the histogram. [3 marks]
d) Use the histogram to complete the table. [3 marks]